Keeping **Chickens**

Keeping **Chickens**

The Essential Guide for First-time Keepers

DAVID SQUIRE

APPLE

First published in the UK in 2012 by
Apple Press
7 Greenland Street
London NW1 0ND
United Kingdom
www.apple-press.com

ISBN 978-1-84543-444-1

Text copyright © David Squire 2012

of David Squire to be identified as author of this work
erted by him in accordance with the Copyright, Designs
and Patents Act 1988.

1 3 5 7 9 10 8 6 4 2

PRODUCED BY
Fine Folio Publishing Limited
6 Bourne Terrace, Bourne Hill, Wherstead, Ipswich, Suffolk, IP2 8NG, UK

DESIGNER
Glyn Bridgewater

ILLUSTRATOR
Coral Mula

EDITOR
Alison Copland

Printed in China by Voion.

Contents

Introduction

Allowing chickens into your life is a major decision as they require attention every day throughout the year. They entirely depend on you for their well-being. However, the rewards are high and not only will your dedicated team of egg-layers provide you with fresh eggs, but their lives are full of interest and will give you endless amusement. Some chickens can even be kept as pets and, it is claimed, a few breeds even answer to their names! Even if they do not respond, your tone of voice calms them.

By keeping chickens in your back garden or as free-rangers, you will know they are having as near-natural lives as possible – not confined to small wire cages. Providing hens with a relaxed, non-stressful existence is important as this encourages the laying of top-quality and wholesome eggs. Remember that 'Happy hens lay the best and most eggs'. Additionally, if you decide to raise a few chicks and are fortunate enough to see a chick pecking its way out of an egg, it is an experience you and your family will never forget.

Keeping chickens free from predators, including foxes and rats, is a constant problem as the ingenuity of vermin is amazing in their quest to attack your birds. Strong and high fences help to deter them.

This book enables you to get to know chickens, from their history and development to types of combs, feather markings, life expectancy and sexual habits. The range of light and heavy large-fowl breeds, as well as bantam-sized types, is described in detail. There is also information about breeds specifically kept for egg-laying, for meat, or for both.

Choosing the best way to keep your chickens is fully described, including the space they need, correct siting and equipment. The day-to-day care chickens require is also detailed and encompasses feeding, watering, cleaning out and temperature considerations from winter to summer. Also described are ways to catch and inspect chickens, handling them and the things to check to ensure they remain healthy.

How to encourage hens to lay eggs is thoroughly explained, from what an egg is to egg-laying cycles and how many eggs you can reasonably expect each year. Collecting and storing eggs is not neglected, together with advice about keeping a cock-bird in with your hens.

Raising chickens for their meat is important, and how to look after them, suitable breeds, plucking, drawing and trussing are detailed.

Raising your own chickens has great appeal and this comprehensive book includes information about hen and cock-bird characteristics for breeding, replacing aged hens, incubation, hatching and rearing.

Inevitably, pests, diseases and problem habits occur in chickens and these are major considerations – they are described in detail, together with ways to prevent them.

Keeping chickens is a healthy, fun and happy way to use your spare time throughout the year. It is also a way to ensure chickens are kept in a natural and caring environment.

Happy chicken keeping!

These Redcap chickens have large and distinctive combs that resemble caps.

GETTING TO KNOW CHICKENS

Origin of the species

There are more chickens on this planet than any other species of bird, with estimates of 28 billion at any one time. They are a primary source of food in many parts of the world and are commercially and domestically kept for their meat and their eggs. The range of breeds is extensive and both large-fowl breeds and bantams are described in detail on pages 28–61.

Male birds often make their presence known in the early morning with a 'cock-a-doodle-doo'.

Where did chickens originate?

The ancestry of modern-day chickens was earlier thought to be from both the Red Junglefowl (*Gallus gallus*) and Grey Junglefowl (*Gallus sonneratii*), some 10,000 or more years ago. However, recent genetic studies suggest that the Grey Junglefowl alone is the most likely ancestor.

Early domestication of chickens was first thought to have taken place in India, but Southeast Asia and probably Vietnam are currently considered to be the most likely areas for the beginning of what is now a global food industry.

From India, domestication spread to Asia Minor, then Greece about 7,000 years ago. This led to their introduction to Egypt during the 18th Dynasty (1550–1292 BCE). Chickens spread from Thailand to China by 5,000 BCE, and to Japan more than 2,000 years ago.

Chickens spread by land and water into Europe, with Julius Caesar discovering them established in Britain in 55 BCE. The British, who began documenting chickens before the 1500s, helped in their spread to other parts of the world, including to Australia and North America.

It is now possible to find chickens in most countries throughout the world, where they are highly prized for their eggs and meat. Indeed, if it were not for chickens the diets of many people would be radically worse.

Few farmyard sights are as idyllic as a mother hen with her chicks tightly clustered around her.

Physical appearance

All chickens have basically the same appearance, although this is influenced by the colours and patterns of their combs (see page 20), feathers (see page 21), and sizes (see 'Range of chicken breeds', page 28, and 'Range of bantam breeds', page 34). Additionally, there are clear differences and attitudes between male and female birds. Usually, male birds are more aggressive than females.

Cock-birds usually have more flamboyant tail feathers and larger combs than hens.

Hens are usually smaller and weigh less than cock-birds and tend to be dominated by them.

A cock-bird likes to make his presence known, by both ushering and guarding his hens.

General characteristics of hens and cock-birds

• **Cock-birds**: Larger than hens and differentiated by their handsome plumage. They often have long, flowing tails and shiny, pointed feathers on their necks. Additionally, they have large, handsome fleshy combs and, often, spurs on their legs.

Often, a cock-bird has a dominant strut which he uses to impose authority over female birds. But perhaps his most well-known characteristic is an ability to produce a 'cock-a-doodle-doo' noise, which many people find disturbing and irritating, especially early in the morning! This is a territorial signal to other cock-birds to keep away from both him and his hens. Additionally, cock-birds become noisy when threatened by predators, becoming ferocious and fearless opponents.

The presence of a cock-bird in with your hens is not necessary for the production of eggs for eating. They are, however, essential if you want to breed your own chickens (see 'Raising your own chickens', page 140).

Cock-birds are protective about their ladies and chicks and may reveal unexpected largess; when he finds food he often encourages his hens to eat first. He performs this by clucking in a high pitch, at the same time picking up and dropping food. This behaviour can also be seen in mother hens when encouraging chicks to eat.

• **Hens**: Smaller and lighter than males, with combs and tail feathers usually less flamboyant. They lay eggs; these are mainly for human consumption but also, when fertilized by a cock-bird, for producing chicks.

• **Flight**: Domestic chickens are not capable of long-distance flight, although light breeds are able to flutter and jump for short distances. Some breeds are able to leap and fly into the low branches of trees.

• **Gregarious nature**: Hens prefer to live in groups and, in a back yard or garden situation, one formed of three to six is about right to cater for their social activities. A hen on her own will be lonely and restless and see you as her main companion – and, possibly, leader of the pack.

Gender terminology

★ **Young males and females**: Widely known as chicks.

★ **Females**: Young females are pullets; over the age of about one year they become hens.

★ **Males**: Until he is about one year old, a male bird is known as a cockerel. Then he becomes a cock, cock-bird or rooster. These terms vary from one country to another and rooster is very popular in North America.

★ **Castrated males**: Known as capons (see page 132).

What you need to consider

Keeping chickens in your garden or on an allotment is surprisingly straightforward, with few difficulties. However, here are some of the things you will need to consider.

• **Provided you intend to keep fewer than 50 chickens** there is no need to apply for permission or to register them with any authority. Ensure your chicken keeping is not commercial.

• **If you wish to keep more than 50 chickens** it is considered to be a commercial venture and must be registered with the authorities in your country.

• **Check with the deeds of your property** to ensure you are legally able to keep chickens. If there is any doubt about this, consult a legal adviser.

• **Consult with your local authority** to check if there is a bye-law in your area that prevents you keeping chickens in your garden. This will prevent problems arising later.

A cock-bird acting with determination and authority to maintain order in his group.

- **Essentially, talk with your neighbours** to see if they are happy about living next door to a group of chickens. This is very much a matter of getting their approval and goodwill, which will help to prevent disagreements occurring later.

Don't hesitate to invite them round when your chickens are established, so that they can see at first-hand what a pleasure it is to keep chickens – and a gift of a few fresh eggs usually helps to seal their approval.

Household pets and chickens

When your chickens first arrive and you get them settled into a chicken house or enclosure, your family pets will wander over and give them some attention. If you can be present at the same time, to talk to them and stroke them, they will not feel threatened by the 'new boys on the block'.

- **The desire of a dog to be friendly** depends greatly on its disposition and breed – some are passive and will just wag their tails at the sight of chickens. Others may be too inquisitive. Therefore, keep a check on their attitude.

- **Chickens kept in wire-netting enclosures** (perhaps formed of a hen house and scratching run) will not be at risk, but those living as free-rangers could be in danger.

- **Cats are inquisitive** but after a few inspections usually ignore chickens. Young chicks will be at risk but when accompanied by 'mum' are usually safe; a hen while looking after her youngsters is a formidable opponent. Your own cat may, eventually, accept the chickens and just ignore them, but local neighbourhood cats can be a problem.

- **Dogs are sometimes tempted to attack lone chickens**, chasing and stressing them. Such chickens – if they survive – may cease laying. If a dog kills a chicken, make it sit in front of you and hold the dead chicken near its face. Scold the dog in a harsh voice. Occasionally, a dog (and this depends on the breed) will act as a protector, herding the chickens and, if necessary, barking to raise the alarm if predators are a threat.

Cats will eventually ignore cock-birds and hens, but pose a problem for chicks when they are on their own.

Vagrant chickens

It is inevitable that one of your chickens will eventually escape and trespass on neighbouring properties. This invariably causes problems because chickens eat and damage plants. Also, they scratch and disturb soil, causing problems in vegetable and flower borders newly sown with seeds. In law, you are responsible for correcting the problem as well as offering compensation.

Litigation is very expensive, and therefore regularly checking fences and gates is essential to prevent hens escaping and causing problems.

If your hens wander on to a road and are killed there is no liability on the part of the motorist to pay compensation. Additionally, if it is proven that you did not take necessary precautions to prevent their escape, you may be liable for any resulting accident.

Understand their pecking order

Sometimes known as the 'peck order', this now popular term describes the social structure and hierarchy within groups of animals – those that are dominant, those that are submissive, and the ones in between. Establishing a pecking order is essential for the welfare of a group of birds because it helps to reduce discord within a group.

The Norwegian zoologist Thorleif Schjelderup-Ebbe (1894–1982) wrote about the pecking order of chickens in his PhD dissertation in 1921, partly based on observations recorded by him since the age of ten on his own chickens.

Development of the pecking order in chickens starts early in their lives. For example:

• **From as young as two weeks** chicks playfully peck at each other. This tentative and experimental activity soon makes some chicks wary and others confident.

• **When nearing puberty** (12–16 weeks for cockerels and 10–12 weeks for pullets) pecking intensifies in the resolution of disputes – sometimes aggressively.

• **At about 26 weeks of age**, the eventual pecking order is established and the birds tend to settle down.

• **If there is only one cock-bird** within a flock, he becomes dominant and the top animal.

• **At this stage, hens start to compete** with one another for the cock-bird's attention, often aggressively and with a determined desire to diminish the roles of other hens, which they consider to be competitors and rivals.

• **If other cockerels are introduced** into the group, mayhem results. Quarrelling and fighting becomes common between males, often spreading to the entire group.

• **Inevitably the two most powerful roosters** fight, vying for top position and dominance over the hens.

• **Cockerels at the lower level of dominance** often become shunned and may be pushed out of the group.

• **Where hens are in a group** and do not have a cockerel with them, they battle to determine the most dominant hen. Interestingly, the top hen may try to crow and even to sexually mount other hens. A hen with a large comb often has the best chance of being top hen.

Looking for grubs and vagrant insects – with heads down – is a popular activity for chickens.

- **Where squabbles occur between hens**, do not use your hands to separate them. Rather gently spray them with clean water. When in full battle, hens are dangerous opponents.

- **Should the squabbling continue**, remove and rehome the weaker one; if the stronger and more dominant one is removed, squabbling inevitably continues among the group. Also, you will have difficulty in introducing the dominant one to another group of hens.

- **When a cockerel is introduced** into a group of hens where one of them is established as top hen, he will be resisted and, often, aggressively attacked. He will never be allowed to mount a hen and probably become a sad and dejected fellow.

- **Introducing new hens into an established group** can be difficult as existing birds may not wish to accept them. Two ways offered in chicken folklore to overcome this problem are to wait until night, when it is pitch dark, or to rub garlic over the hen that is about to be introduced. However, take care to use the latter method reservedly as the odour lingers.

The pecking order and you!

Chickens soon get to know their feeding routine and often consider the person who dispenses food to be the leader of their pack, someone to revere. Depending on their gender, they will approach you in different ways. For example:

- **Cock-birds**: Until they get to know you well, they invariably approach you cautiously. When in a group of hens, a cock-bird strongly believes he is the leader and protector of his 'ladies'.

He will approach you, with direct eye contact, and give a slow flap of his wings. The spreading of wings makes him appear larger and therefore intimidating and defensive.

- **Hens**: These may greet and approach you in the same way as cock-birds – looking at you and spreading and flapping their wings. Eventually, after getting to know you well, a hen will be friendly and may stand on your shoes and peck at your trousers. This usually indicates she wants to be picked up, caressed and talk to. However, giving one chicken extra attention may lead to jealousy among the other hens.

Colours in chickens

The spectrum of colour produced by a chicken's feathers is impressive and greatly admired. These range from white to black and include buff, red and blue, as well as shades of yellow. Many established breeds have colour variations and this has resulted in an even wider range of colour; this has generated a colour vocabulary, with some terms specific to chickens, and others more general.

- **Bay**: Warm, golden-brown, especially when highlighted by sunshine. Even in winter it has a warm richness.

- **Birchen**: Mixture of black and pure white; probably an allusion to the Silver Birch tree which has silvery-white bark and irregular dark fissures.

- **Black**: Lustrous greenish-black, fully revealed when in strong sunlight. It is certainly a handsome colour.

- **Black-breasted red**: Refers to birds which are mostly black but with red hackles, shoulders, back and parts of their wings.

A rich, tawny-brown appears to be in harmony with nature.

White feathers create colour contrasts for combs.

- **Blue**: Usually defined as slate-blue, but this can vary and include lavender. It is a very restful colour.

- **Brown-red**: Fusion of black and orange.

- **Buff**: Fusion of orange and yellow, with a rich golden shade.

- **Chestnut**: Dark, reddish-brown.

- **Cinnamon**: Dark, reddish-buff.

- **Duckwing**: Having a distinct bar across the wing of a male bird. Such birds are easily recognizable.

- **Fawn**: Light, brownish-tan.

- **Horn**: Dark brown.

- **Mahogany**: Deep, reddish-brown.

- **Partridge**: Patchwork – resembling the colours on partridges.

- **Porcelain**: A soft and gentle fusion, created by straw-coloured feathers tipped in white and with a pale blue stripe through part of the feather.

- **Quail**: A colour pattern similar to that of quails. Basically, this is when black neck, back and saddle feathers are laced in golden-bay.

- **Red**: A dramatic colour; it may vary and includes mahogany-red and rich dark red.

- **Salmon**: Less dominant than red and seen as a reddish- or pinkish-buff; said to resemble the colour of cooked salmon.

- **Self-coloured**: Refers to a chicken having just one colour.

- **Silver-pencilled**: A colour pattern formed of silvery-white and pencilled feathers.

- **Slate**: Dark, bluish-slate. Sometimes it may appear to be very nearly black.

- **Wheaten**: This is the colour of wheat when glowing in sunlight – yellowish ochre.

- **White**: Feathers entirely white.

- **Willow**: Yellowish-green.

Some feathers have a handsome, reflective gloss.

Tolbunt tricolor colouration in a Polish breed.

Range of combs

Every ordinary chicken has a comb on its head. The only exceptions are exhibition breeds and game fowl that have been dubbed for show (see Glossary entry for 'Dubbed').

★ **Combs are nature's way** of enabling chickens to identify each other and for males to attract females.

★ **The shapes and sizes** of combs vary widely from one breed to another, with those on male birds being larger and usually more flamboyant than those on females. Part of its role is to attract hens.

★ **Combs are fleshy and soft** to the touch.

★ **Because blood is able to flow** freely through a comb and does not have the cold-weather protection provided by a covering of feathers, in some breeds they are vulnerable to damage. During extremely cold weather, they can become black. However, a coating of petroleum jelly gives them some winter weather protection.

★ **Combs have a functional role** in helping to keep a chicken cool during hot weather.

★ **Most combs are red**, but there are a few exceptions. Silkies and Sumatras have purple combs and Sebrights have purplish-red ones.

Types of combs

Combs have been classified and grouped into several distinctive shapes. These are:

● **Buttercup**: Beautiful cup-shaped comb that starts in a single ridge at the top of the beak. It then creates a circle formed of

Combs are remarkable head features that assist in breed identification (see below and opposite).

regularly spaced points that resemble those on antlers. Few breeds have this type of comb.

● **Horn**: Also known as the 'V-shaped comb', this has two raised and spiky pieces that look like horns. They are joined at their bases.

● **Pea**: A low, medium-sized comb formed of three ridges running lengthwise from the top of the beak to the top of the head. The middle ridge is slightly higher and larger than the other two. Additionally, it has slight serrations or undulations.

● **Rose**: Broad, solid but fleshy, the comb lies almost flat on top of the head. It is tube-shaped, with small, rounded bumps extending from the top of the beak to the back of the head. It ends in a single spike (leader) at the back of the head.

- **Single**: This is the most common type of comb, formed as a thin, single line starting at the top of the beak and extending to the top of the skull. On top there are five or six deep serrations that create several sharp points. Combs on male birds are always erect, while those on females often fall sideways, depending on the breed.

- **Strawberry**: A low and distinctive comb, known as a strawberry because of its shape and rough surface.

- **Walnut**: Distinctive, almost round comb, slightly wider than it is long. It reveals a narrow, cross-way indentation slightly to its front. Silkies have combs with this shape.

Feathers and their markings

Often, the first noticeable feature on a chicken is its colour or range of feather colours. Some colour styles are sex originated and appear only in female or male birds of a specific breed.

Feathers grow from the outer skin layer and form a bird's plumage. They give protection from low temperatures, excessive warmth and rain; they also readily identify a breed.

Each feather has a central, hard quill (known as a shaft) with barbs attached to it that stick together and form a smooth surface. Colour, size and shape of feathers vary and include:

- **Barred**: Distinctive, with feathers showing horizontal stripes in two different colours. Sometimes, the cross-bars are evenly sized and shaped, while in some breeds they are irregular.

- **Cuckoo**: Irregular light and dark bars with the ends of feathers being dark.

- **Laced**: Beautiful, dainty and distinctive, with each feather having a narrow border colour around the outside. The range of colours is wide: some silver, others white, buff or golden.

- **Mottled**: Feathers have white tips. However, not all feathers on a bird may have this feature.

- **Pencilled**: Distinctive bars, some thin, in rows that encircle a feather. Usually, these are mostly seen in female birds.

- **Spangled**: Tips of feathers with black or white markings, often teardrop-shaped, although with some breeds this is more sharply edged.

- **Stippled**: Feathers display dots in a colour that creates an attractive contrast to their background.

- **Stripped**: Uniform colour along a feather's centre and surrounded by lacing.

Sebright bantams have distinctive, black-edged feathers.

Chickens' senses and sounds

Chickens are now known to have excellent colour vision that is certainly better than that of human eyes. Each chicken has an eye on either side of its head, enabling it to see things in three dimensions as well as having all-round vision. A chicken focuses its left eye first, then the right; this explains why head movement is rapid and from side to side. As they walk, they continually move their heads.

Recent research confirms their excellent colour vision. The human retina has light receptors sensitive to red, blue and green wavelengths, whereas a chicken has retinas also able to detect violet wavelengths, including some ultraviolet. In addition, it has receptors known as double-cones that help in the detection of motion, enabling chickens to be extra alert to possible danger.

This bonus of superb colour vision is thought to enable them to detect colourful fruits and other food more easily, as well as aiding in their quests to find a mate.

Birds are descendants of dinosaurs and during the 'age of dinosaurs' – which ended some 65 million years ago – early mammals hid during the day and scurried around in the undergrowth at night, when they were less likely to become a meal for dinosaurs.

It is therefore suggested that chickens – as well as other birds – owe their good daylight vision to not having spent a long evolutionary period in the dark.

A chicken's hearing

Chickens have a well-developed sense of hearing. This fact may appear questionable because they do not have visible ears. Instead, their auditory canal is hidden behind feathers and a flap of skin. Nevertheless, chickens are well aware of ambient sound.

● **A chicken's ability to hear well** should not be surprising as both courtship and successfully remaining in a flock rely on being able to recognize both friendly and warning sounds. Additionally, even when a chick is still in its egg it identifies through sound with mum, receiving comfort and information.

● **Birds lack the external and easily visible feature** that so quickly identifies an ear. Their ears resemble the three basic parts found in a human ear. The outer ear is a tube leading to the ear drum (tympanum). The middle ear has a single bone stretched across it, called the columella. The inner ear is bathed in fluid; the outer and middle parts are filled with air.

● **Chickens, as well as other birds,** hear things differently from humans. Birds remember and recognize sound in something similar to 'perfect' pitch, whereas humans receive sound in 'relative' pitch. In practice, this means that if humans receive sounds in one octave they are able to recognize it in a different octave. Chickens cannot do this, although they can recognize a fundamental note combined with harmonies. This enables them to respond to sound – and imitate it – in many ways.

Chicken speak!

If you have sat, watched and listened to chickens – whether in a poultry house, walking about a run or foraging in a field – you are soon aware they are able to communicate and relate to each other through sound.

• **A hen when about to lay an egg** produces a laboured, almost complaining note. After laying an egg she resorts to a cackle of pride and contentment.

• **Cock-birds are famed** for their 'cock-a-doodle-do' and are able to do this from the age of three to four months. They cannot instantly achieve this call, however, and start a few weeks earlier with a few, rather subdued, squeaks.

• **The volume of a 'cock-a-doodle-do'** is influenced by the size and bodyweight of the bird.

• **Angered and threatened cock-birds** produce a low-pitch sound to warn their hens of approaching danger.

• **A cock-bird when trying to attract** his hens emits a high-pitched but friendly and non-aggressive note to capture their attention and draw them to him.

This handsome cock-bird radiates self-importance and confidence in his endeavours to attract hens.

• **Hens and their chicks relate** in various ways. Hens cluck, while chicks cheep, and if they become separated from mum quickly break into a high-pitched tone. Indeed, the higher a chick's tone, the more rapidly mum responds.

• **Contented hens cluck slowly** and with an even tone, but if alarmed produce a gagaga sound.

Foraging and eating techniques

Eating and foraging both form an essential part of a chicken's life and growth. Because the birds randomly pick up grains of food from the ground, good eyesight is essential (see page 22).

★ A chicken can identify a grain of food from a distance of about 90 cm (3 ft); when its head is 50–75 cm (2–3 in) from the food, it leans down and eats it.

★ Because a chicken first focuses with its left eye, then the right, this explains why there is rapid head movement, from side to side, before focusing on a grain of food.

★ In addition to grains of foraged food, worms and bugs are on the menu.

Sexual activities

Chickens believe in 'free love' and mate throughout the year – with any partner! Attractive combs and feathering play a role in attracting a mate, but many poultry experts report that some cock-birds will mate with any hen that does not move fast enough...

Cock-birds perform a ritual when endeavouring to interest a hen in sexual activity (see right).

- **Chickens mate throughout the day**, but are more sexually active in the afternoon as it increases the chance of producing a fertilized egg.

- **The wooing part of chicken sex** is a cock-bird standing upright in front of a hen and fluffing out his neck feathers. He then dances around her, at the same time extending and lowering the wing closest to her. This sexual overture is often known as the 'wing-fluttering' or 'waltz' stage.

- **The sexual act** is when a cock-bird pounces, jumping on the back of a hen (at the same time using his wings to maintain balance). He also pecks at the hen's neck to retain balance and position; using his feet (sometimes known as 'treading'), he pushes her down. The hen lowers her head, raises her tail and the cock-bird presses his cloaca against hers; at the same time, he releases sperm. This may last about ten seconds and is known as the 'cloacal kiss'. The sperm then ascends into the hen's oviduct and fertilizes the egg.

- **After his sexual performance**, a cock-bird moves away, sometimes circling the hen and boastfully crowing.

- **After her sexual activities**, a hen usually stands, shakes her feathers and moves away.

- **Hens reluctant to indulge in sexual activities** escape by running away, by moving their tails from side to side, or just by crouching.

The broody hen

This is a natural and inherited function to ensure the development of the next generation of chickens. Some breeds have hens that are particularly likely to become broody and therefore be good mothers.

★ **Ancestors of modern chickens** laid two clutches of eggs each year, one in spring and the other in late summer. Each clutch was formed of 12–15 eggs. This characteristic has been bred out of many chickens, especially those raised for commercial egg production. Nevertheless, some older and heavy breeds, as well as bantams, still possess this characteristic.

★ **While sitting on a clutch of eggs,** a hen both defends them and encourages hatching (see page 148 for detailed information).

★ **Broodiness can be discouraged** by putting a broody hen in a slatted coop. Basically, this is a wooden box, 75–90 cm (2–3 ft) long, 60 cm (2 ft) in depth and 60 cm (2 ft) high, with a sloping, waterproof roof. The floor is closely slatted and with a widely slatted door at the front that can be raised or lowered. The broody hen is placed inside and able to put her head between the slats at the front to look out – but not to escape. Water and food are freely available to her.

The coolness of this coop, together with the sight of other hens walking about and enjoying their freedom, usually discourages her from continuing with her broodiness. After a few days in the coop, she can be put back with the other hens; she usually then starts to lay further eggs.

A broody hen needs a quiet, warm and cosy area, free from intruders and vermin.

Rescued hens

Often known as 'ex-bat' or 'spent' hens, these are fowls which have lived in battery cages and have reached the end of their commercially economic lives producing eggs.

At the end of their commercial life, generally around 72 weeks old, the hens are sent for slaughter where the meat generally ends up in lower grade products such as pet food and cheaper fillings. Alternatively, they are offered for rehoming through welfare organisations.

Rehomed hens can go on to become much loved and individually named family pets, with well over 300,000 having been homed in the last five years across the UK. Many people find their naturally docile and inquisitive nature makes them ideal family and first-time chicken keeper pets, and will usually go on to lay a good number of eggs for their new owners.

While they generally come out of the commercial sector looking 'tatty', and are unfit, they are healthy, as it is not in the farmers' commercial interests to keep unwell birds that do not

Rescued hens quickly adjust to a life outside and go on to become much-loved pets.

produce eggs. Within a few weeks they will regain their fitness and natural instinctive behaviour and feather regrowth will quickly follow on.

Most welfare organisations home a minimum of three hens (in each new 'home'), with the largest being the British Hen Welfare Trust, which works in partnership with UK farmers and campaigns to raise awareness of higher welfare eggs and the use of caged eggs 'hidden' in processed food. If you would like to find out more, or rehome some hens, visit their website at www.bhwt.org.uk for details.

Rows of chickens in soulless cages have little appeal for many chicken enthusiasts and humanitarians.

While the pressure of intensive laying during their commercial lives do take a toll on their bodies, rescued hens usually go on to lay enough eggs to keep the average family self-sufficient.

RANGE OF CHICKEN BREEDS

Light or heavy large-fowl breeds

A 'chicken is a chicken', but is it? The range of individual breeds throughout the world is wide and varied, in size, shape, colour and temperament. Additionally, with many breeds there are several colour variations, from clear and distinctive single colours to those that display a range of shades in their feathering.

Range of breeds

As early chickens spread throughout Asia and into Europe and other parts of the world, so individual breeds were developed. Most countries have their favourite breeds of chickens, which have been created to suit local climates and demands for egg or meat production – or both.

The majority of breeds are classified as 'light or heavy large-fowls'. Additionally, there are small breeds which became known as 'bantams'. Some bantams are classified as 'true' bantams because they are naturally small and do not have large counterparts. Whereas other bantam breeds have arisen because chicken breeders created diminutive versions of ordinary-sized chickens. A range of bantams is described on pages 54–61.

A–Z of light and heavy large-fowl breeds

The 'heavy' breeds, as one would suspect, are heavier in bodyweight than 'light' breeds and because of their larger size they eat more. They are also less likely to fly and therefore can be kept in runs with lower fences. Heavy breeds need more space in their enclosures, as well as at night when locked away in chicken houses.

Light breeds sometimes have a flighty, skittish and nervous disposition, although they often lay as many or more eggs than some heavy breeds (which may primarily have been bred for their meat, rather than egg-laying qualities). Interestingly, the majority of light breeds originated in Mediterranean and northern European regions.

Some breeds of chicken have an aloof dignity, which is far from their dinosaurian ancestry!

With many light or heavy large-fowl breeds there are varieties that have colour variations in their plumage. This especially applies to heavy breeds, although some light ones, such as Silkies, have distinctively coloured variations. The breeds described in this chapter are individually detailed under the most popular form, with colour variants explained in the description. Others, where there is just a single form, are headed under the main name.

Ancona

This light, soft-feathered breed should not be confused with the Leghorn (see page 43), another breed originating in Italy. The Ancona is primarily kept for its egg-laying qualities. Its feathers are a fusion of dark beetle-green and black, with feathers distinctively tipped in white and rounded at their tips. This creates an attractive downward pattern. The legs are yellow, with green-black mottling.

There are two comb forms: Single Comb (developed in 1898) and Rose Comb (originated in 1914). They have white earlobes and bay-coloured eyes.

It is best kept as a forager; also happy when in a run.

A Mediterranean breed, developed in Italy and named after the province of Ancona on the Adriatic coast. During the late 1840s examples of this breed were introduced into England and shown at the Great Exhibition in London in 1851. In 1888, specimens were taken from England to North America.

Colour variations: None.

Eggs: 180–240 each year.

Egg colour: White to cream.

Weight: Hen 2 kg (4 lb): Cock 2.7 kg (6 lb).

Temperament: Friendly, alert and sometimes flighty – needs to be handled from a young age.

Ancona chickens are distinctive, with attractive combs.

Many Andalusian chickens have beautiful blue plumage.

Andalusian

Usually grouped as a Mediterranean breed and originating in Andalusia, it is a rare light breed. There is only one colour, a light shade of blue – they do not always breed true (see below). It has a single comb and long, dark slate-blue shanks and toes.

When breeding from a blue male and a blue female, the resulting offspring are 50 per cent blue, 25 per cent black, and 25 per cent splash (silver-white, with splashes of blue). However, it is a matter of genetics, and when breeding from a black and splash this usually produces 100 per cent blues (but even this is uncertain unless a blue-bred black is used). The Father of Genetics, Gregor Mendel (1822–1884), included these chickens in genetic experiments.

Colour variations: None.

Eggs: 160–180 each year.

Egg colour: White.

Weight: Hen 2.3–2.7 kg (5–6 lb): Cock 3.2–3.6 kg (7–8 lb).

Temperament: Active yet docile – ideal as family pets but must be handled from when young.

Appenzeller Spitzhauben chickens have fluffy, upright crests.

Appenzeller Spitzhauben

Distinctive light breed, which originated in the Appenzell region in northeast Switzerland during the 16th century and is recognized as the national chicken breed of Switzerland.

It is sometimes kept for its ornamental appearance, although it does provide plenty of eggs. The word Spitzhauben means 'pointed hood' and this is reflected in the forward-pointing crest, which is a specific breed feature present in both males and females. It is a style claimed to derive from the fluffy, fancy hats women wore in that region.

It is most usually seen in a silver spangled form, but also officially recognized are the Gold Spangled and Black colour varieties. They have dark brown eyes and dark beaks, while the shanks and toes are blue.

Colour variations: There are three main varieties (as above), although blue and chamois Spangled variations are known.
Eggs: 160–200 each year.
Egg colour: White.
Weight: Hen 1.35–1.6 kg (3–3.5 lb): Cock 1.6–2 kg (3.5–4.5 lb).
Temperament: Active and superb at foraging. Tends to be flighty, although will calm easily with careful handling.

Araucana

A light, soft-feathered breed known in North America as the South American Rumpless because of its absence of tail feathers. Its ancestors originated in northern Chile and it was introduced into Europe in the early 1900s.

These chickens were first raised by Araucana Indians in the Arauca region of Central Chile.
Colour variations: The American Poultry Club recognizes five colour variations – Black, White, Black-breasted Red, Silver Duckwing and Golden Duckwing. Slightly differently, the Poultry Club of Great Britain recognizes four types – Large-tailed, Bantam-tailed, Large Rumpless and Bantam Rumpless.
Eggs: 150–180 each year; sometimes more.
Egg colour: Blue – an even colour throughout the shell. Sometimes they are a greenish-blue.
Weight: Hen 2.3–2.7 kg (5–6 lb): Cock 2.7–3.2 kg (6–7 lb).
Temperament: Fast-growing chicks that rapidly develop into active and alert adults. Hens are sometimes subject to broodiness, which makes them excellent mothers.

The Araucana breed originated in South America.

Ardenner chickens are active and need plenty of space.

Ardenner

An old and light breed, originally from the French-speaking area of the Ardennes in Southern Belgian. It is rare, handsome, and possesses good egg-laying qualities.

The breed is claimed to be descended from the Gauloise, an ancient Gallic type; the current Ardenner reveals an elegant and upright bearing. The breed is also characterised by pigmented facial skin, ranging from dark red to purple, an upright comb and long tail held erect. However, a rumpless form (without erect tail feathers) is also known. The eyes, beak, shanks and toes (and nails) are darkly coloured.

The Ardenner is best kept as a free-ranger as it is superb at digging up bugs and snails. However, its flight abilities, like many other light breeds, make it necessary to have a high fence around its enclosure.

There are three basic colour forms, black-red, birchen, and yellow-birchen (others are listed below).

A bantam-sized form of the Ardenner was recognized in 1904 and is extremely rare, even in Belgium.

Colour variations: Blue-red, silver duckwing, black, golden-necked, and silver-necked silver-salmon.
Eggs: 120-150 each year - sometimes more.
Egg colour: White.
Weight: Hen 2 kg (4.4 lb); Cock bird 2.5 kg (5.5 lb).
Temperament: Cock birds can be feisty but with regular handling are easily tamed and become very friendly.

Barnevelder

A heavy, soft-feathered breed with broad shoulders, wings high and short and rounded breast. The body feathers have a red-brown ground colour, edged in black and with a shiny green sheen. Its neck hackles are black and also have a green sheen. However, those in male birds have a red-brown edging. The neat head bears a short, yellow beak with a dark point. Additionally, it has orange eyes and yellow feet and legs.

It has a medley of breeds in its ancestry; between 1850 and 1875, Brahma, Cochin, Malay and Croad Langsham fowls were introduced from Asia to the Netherlands and crossed with local fowl. Subsequently – and slightly prior to 1914 – the Barnevelder breed arose in the Barneveld area of the Netherlands.

Colour variations: The Double-laced form is the most popular, as described above.
Eggs: 180–200 each year.
Egg colour: Brown.
Weight: Hen 2.7–3.2 kg (6–7 lb): Cock 3.2–3.6 kg (7–8 lb).
Temperament: Good-natured, alert and friendly. However, it tends to be lazy and therefore is best kept as a free-ranger. Hens become broody very easily and make good mothers.

Barnevelder chickens have many breeds in their ancestry.

Barred Rock

This is the best known and most widely kept variety of the Plymouth Rock, a heavy breed (see page 46) and one of the most popular chickens in North America. It is ideal for both producing eggs and eating. It is a heavy, soft-feathered breed with a deep, long and broad, plump body. The barring is fine and distinctive, with each feather having well-defined black and white horizontal bars. Young chicks are dark grey to black, with some white patches on both the body and head.

   Its origination dates back to the 1800s in New England (USA) and initially originates from crosses between Black Javas and Dominiques, subsequently crossing the offspring with further Dominiques.

Colour variations: There are six other varieties of the Plymouth Rock breed and these are Blue, Buff, Columbian, Silver-pencilled, White, and Partridge.

Eggs: 220 each year – sometimes more.

Egg colour: Light to medium brown, with rich yellow yolks.

Weight: Hen 3.4 kg (7 lb): Cock 3.6 kg (8 lb).

Temperament: Well-mannered, quiet and docile, making it an excellent pet for children. Its egg-laying ability is prolific and undiminished by cold winter weather.

The Barred Rock has distinctively coloured plumage.

Black Australorp

This heavy, soft-feathered breed has glossy-black feathers with a deep, lustrous-green sheen. Sometimes, it is just known as Australorp. Its face is unfeathered; it has white skin and pinkish-white soles to its feet, with similarly coloured toes. It is popular for laying eggs and when its egg-laying life is over makes a superb stewing chicken.

   It was developed in Australia from Black Orpingtons and introduced into Britain in 1921. The name Australorp is sometimes claimed to be an abbreviation for Australian Black Orpington.

Colour variations: None.

Eggs: 200 each year – sometimes more.

Egg colour: Tinted brown.

Weight: Hen 2.9–3.6 kg (6–8 lb): Cock 3.4–4.5 kg (7–10 lb).

Temperament: Active but placid and friendly.

The Black Australorp has shiny plumage.

The Black Silky reveals a 'furry' appearance.

Black Silkie

In Europe, this breed is classified as a large-fowl light breed, whereas in North America it is mainly known as a bantam. However, both conceptions of this breed have soft feathers. They have five toes on each foot, dark wattles, blue earlobes and walnut-type combs. Additionally, Black Silkies appear in two forms – Bearded and Non-bearded. The Bearded types have a distinctive extra muff of feathers under the beak area and extending to the earlobes. They are unable to fly and therefore can be kept in areas with low fencing.

Silkies have a long and amusing heritage: during his travels in China in the 13th century the Venetian traveller Marco Polo (c.1254–1324) recorded them as a 'furry' chicken. Later, in 1599, the Italian naturalist Ulisse Aldrovandi described them as 'wool-bearing' chickens and 'resembling black cats'. Silkies arrived in England in the mid-1800s.

Colour variations: In addition to the Black Silkie, there are White, Gold, Blue, and Partridge varieties.

Eggs: 80–100 each year.

Egg colour: Tinted or cream.

Weight: Hen 1.4 kg (3 lb): Cock 1.8 kg (4 lb)

Temperament: Friendly, lively and ideal as pets for children. Unfortunately, they are poor layers, but they compensate for this by being good mothers and will incubate their own eggs as well as those from other chickens.

The Black Rock is a handsome hybrid.

Black Rock

This hybrid has become very popular and has handsome, all-weather, dense plumage, predominantly black and with varying amounts of chestnut from its chest to neck.

The development of hybrids takes many years, with details of the breeds involved often kept highly secret. However, an outline of the Black Rock's development indicates that it was created from selected strains of Rhode Island Red males and Barred Plymouth Rock females.

Colour variations: None.

Eggs: 280 each year – often more.

Egg colour: Brown.

Weight: Hen 3.4 kg (7 lb): Cock 3.6 kg (8 lb).

Temperament: Docile, friendly and not easily stressed. They have a long egg-laying life and are ideal as free-rangers. They seldom need to be de-beaked.

The Meadowsweet Bluebelle™ has distinctive plumage.

Bluebelle

A pretty chicken with blue-grey plumage, ranging from pale grey to dark slate-blue. Invariably these chickens are known as Meadowsweet Bluebelles™.

This increasingly popular breed originated in the Czech Republic and is derived from Maran and Rhode Island Red fowls. It is a hybrid and its offspring revert to type according to the cockerel used to fertilize the eggs. Therefore, always buy fresh hens, rather than trying to raise your own chicks.

Colour variations: None.

Eggs: 250 each year.

Egg colour: Pale brown.

Weight: Hen 3.4 kg (7½ lb): Cock 3.6 kg (8 lb).

Temperament: Docile and easy to handle, making it an ideal pet for children.

Blue-laced Wyandotte

Heavy, soft-feathered breed with a deep chest and rounded rear end. The distinctive feathering has a red-brown ground colour, with feather edges marked in blue.

The Wyandotte range of breeds originated in the USA in the form of the Silver-laced Wyandotte, and the breed was formally recognized in 1883.

Colour variations: Within the Wyandotte range there are many superb varieties and, apart from the Blue-laced Wyandotte, these include:

• Silver-laced Wyandotte: silvery-white feathers with black edges, an effect known as lacing. Its tail is black and its feet are yellow.

• Gold-laced Wyandotte: a beautiful golden appearance, with black edges to its feathers that create a scalloped look.

• Silver Pencilled Wyandotte: less dramatic appearance than some Wyandotte breeds, but nevertheless beautiful and often claimed to resemble a partridge. Hen birds have a silver undercolour, while cock-birds are white with bits of black that graduate to total black in the tail and wings.

Eggs: 200–240 each year.

Egg colour: Light brown to brown (sometimes tinted).

Weight: Hen 2.7 kg (6 lb): Cock 3.8 kg (8 lb).

Temperament: Ideal both as a free-ranger and in a run. Additionally, they often become good family pets.

The Blue-laced Wyandotte is a rich pageant of colour.

The Buff Orpington has been popular for many decades.

Buff Orpington

Popular heavy breed, well known for its egg and meat producing qualities. It has a dense, golden-buff, soft-feathered nature, with a neat head and single comb. Because they are heavy birds, they are usually unable to fly. They are superb mothers.

This breed originated in a village in Kent, England, and was introduced into the poultry world in 1894 by William Cook.

Colour variations: In addition to the Buff Orpington, there are Black, Blue and White varieties. The dark-coloured varieties have dark eyes and legs, whereas paler-shaded ones have red eyes and white legs.

The amount of sun these chickens receive influences their colours; in areas where there is strong sunlight, light shading is essential.

Eggs: 140–160 each year.

Egg colour: Light brown.

Weight: Hen 3.6 kg (8 lb): Cock 4.5 kg (10 lb).

Temperament: Docile nature and ideal as family pets. They are claimed to be able to answer to their names and enjoy being handled.

Buff Sussex

Dual-purpose heavy breed, producing excellent eggs as well as meat for the table. It has a soft-feathered appearance, with a buff-gold, perhaps slightly ginger, body. The head and neck hackles are buff and sharply striped in green-black, while the wings are also buff but with black in their flights. Tail feathers are distinctive green-black.

It has a wide, flat back, deep breast and wide shoulders. Legs and feet are white, and earlobes red. Legs are short, strong and free from feathers. They rarely go broody and are ideal for beginners.

Colour variations: In addition to the Buff Sussex there are several other varieties, including Coronation (developed for the coronation of King Edward VIII in 1936, but this did not happen because he abdicated and his younger brother became King George VI). Most popular and readily available are the White Sussex and Red Sussex. Others include the Light Sussex, Speckled Sussex, Silver Sussex and Brown Sussex.

Eggs: 180–200 each year.

Egg colour: Cream to light brown – and tinted.

Weight: Hen 3.2 kg (7 lb): Cock 4 kg (9 lb).

Temperament: Alert, usually docile and highly adaptable. It is suitable for keeping in scratching runs as well as a free-ranger; it naturally forages for food.

© FIONNA APPLETON/WWW.CHOOKS.CO.NZ

The Buff Sussex is a traditional English breed.

Few other breeds have such glorious plumage as the Buff Orpington.

A proud and handsome Blue Partridge Cochin.

Cochin

Distinctive large and heavy breed with dramatically feathered feet, legs and toes. In the mid-1800s they were known as Ostrich Fowl and thought to be related to buzzards! It is a breed famed for its long, soft plumage that creates a curved body right down to its toes.

The Cochin, as its name suggests, originated in China and was introduced into England in the early 1850s. Earlier, it was known as the Chinese Shanghai (because it was exported from that sea-port city) and Cochin-China. Selective breeding radically changed the Cochin; in the mid-1800s it was highly prized for its ability to produce eggs, often 150 each year and especially throughout winter. However, breed development concentrated on showing it at exhibitions, at the expense of egg production. Therefore, this breed is now best kept for its beauty.

Miniature Cochins: Such is the attractiveness of this breed that miniature versions are popular, in a wide colour range including Barred, Birchen, Black, Blue, Buff, Golden Laced, Partridge, Red, Silver Pencilled, Splash and White. They resemble Pekin Bantams, but do not have such as round and 'tea-cosy-like' appearance. Neither do they have such a distinctive forward tilt to their bodies.

Colour variations: It is known in several colours, including White, Black, Blue, Buff, Cuckoo and Partridge.

Eggs: 10–30 (sometimes more) – each year.

Egg colour: Tinted light brown.

Weight: Hen 5 kg (11 lb): Cock 5.8 kg (13 lb).

Temperament: Friendly, quiet and broody – they make good mothers.

Cream Legbar

This moderately rare breed has a muscular, wedge-shaped body, broad at the shoulders and gradually tapering towards its rear. Its body tends to be flat at its top, with cream-barred neck and saddle hackles, dark grey and tipped with cream. The primary feathers are dark grey and barred, while the secondary ones are grey barred with cream tips. The face is smooth and red, the beak yellow and with pendent white or cream earlobes. The unfeathered feet and legs are yellow, with four toes on each foot.

The Cream Legbar has several breeds in its formation and is a cross between the Brown Leghorn and the Barred Plymouth Rock with the addition of some Araucana blood.

It is known as a autosexing variety, which means that the sex of young chicks can be determined soon after hatching by the colour of the soft down on their heads. Each male has a pale dot on its head and with little or no eye barring. Female chicks have a black or dark brown stripe on their heads. This information enables male chicks to be removed.

Colour variations: Apart from the Cream Legbar, there are Gold Legbars and Silver Legbars.

Eggs: 170 each year.

Egg colour: Blue-green, but sometimes olive.

Weight: Hen 2–2.7 kg (4–6 lb): Cock 2.7–3.4 kg (6–7 lb).

Temperament: Alert, sprightly and inquisitive.

The Cream Legbar is derived from a medley of breeds.

Croad Langshan

A heavy, distinctively black, tall, soft-feathered, large-bodied breed with a long and deep breast. The head often appears to be too small for its body, which has a long, sloping rise to the abundantly feathered tail. This gives the breed a U-shaped appearance, with the head and tail clearly seen at either end. The beak is light to dark horn, while the legs are black and feathered on the outside and the outer toes. It is kept for its egg-producing qualities as well as for the table.

The Langshan, originally from Vietnam, also has its origins in the Langshan District just north of the Yangtse-Kiang River in China.

Colour variations: There are several variations, including the White Langshan and the Blue Langshan.

Eggs: 180 each year – sometimes more.

Egg colour: Brown, with a plum-like bloom.

Weight: Hen 3.2 kg (7 lb): Cock 4 kg (9 lb).

Temperament: It thrives as a free-ranger, as well as in confined situations. The hens are good mothers and the breed is well suited to warm climates. The ability of hens to lay eggs is not diminished in winter.

The Cuckoo Maran originated in France.

Cuckoo Maran

Heavy, soft-feathered breed, ideal for producing eggs as well as meat for the table. Indeed, it was originally kept for providing high-quality meat for the French market as it has a relatively quick-maturing nature. It has a medium-length body, with a good width and depth for meat production. Its tail is held high and the head is neat, with a single comb. The legs are featherless and white, the earlobes red, and the eyes orange-red. This breed tends to broodiness, although it is a variable characteristic in some varieties.

This distinctive and thoroughly commercial breed originated in France and gains its name from the French town of Marans, northeast of La Rochelle in western France.

Colour variations: Apart from the Cuckoo Maran, there are other popular varieties, including the French Copper Maran, French Wheaton Maran, Black Maran, Black Copper Maran, Black-tail Buff Maran, White Maran and Columbian forms.

Eggs: 150–180 each year.

Egg colour: Deep chocolate-brown.

Weight: Hen 3.2 kg (7 lb): Cock 3.6 kg (8 lb).

Temperament: Graceful, docile and dignified, but nevertheless active and responsive. Male birds of some varieties tend to be vicious, but if you are keeping the breed solely for egg production this is not a problem. It does well as a free-ranger and thrives in marshy and rough conditions.

© JONATHAN AYRES

The Croad Langshan is now known throughout the world.

Faverolles

Heavy, soft-feathered breed, ideal for eating as well as producing medium-sized eggs. It has a wide body, round head and reddish-bay eyes. The pinkish legs are sparsely feathered, with feathering mainly on the outer of the five toes. The Salmon variety displays a rich mixture of colours, including white, black and brown.

Faverolles originated in the middle of the 19th century in the village of Faverolles in northern France. Their blood line includes Brahma, Dorking, Crevecoeur, Houdan, Coucou de Rennes and, possibly, the Cochin. By the end of that century the breed was highly popular.

Colour variations: Salmon is the most popular variety. Others include the Black Faverolle, White Blue Faverolle, Cuckoo Faverolle, Buff Faverolle, and Ermine Faverolle.

Eggs: 160 each year – sometimes more.

Egg colour: Light brown or cream-white and tinted.

Weight: 2.9–3.4 kg (6–7 lb): Cock 3.6–4 kg (8–9 lb).

Temperament: It has a gentle and friendly nature, making it an ideal family pet, especially for children. Hens are good mothers, alert and active. It thrives as a free-ranger as well as in an enclosure.

Faverolles have many breeds in their ancestry.

Frizzles are amusing, with an unusual appearance.

Frizzle

Few breeds of chicken are as distinctive as the Frizzle, with its 'straight from under the hair-drier' appearance. It is a heavy breed, usually kept for exhibition purposes but well able to provide eggs for you. It is also kept for its meat. When newly hatched, the chicks have a normal appearance, but quickly the wing feathers develop and turn outwards, creating a frizzled appearance. There are several colour variations (see below).

Charles Darwin (1908–1982) called these birds Frizzle or Caffie Fowls and they are claimed to have originated in Southeast Asia, in Java and the Philippines. When they were introduced into North America, the chicken breeder T. Farrar Rackham of New Jersey played a central role in their development.

Colour variations: These include White, Black, Blue and Buff.

Eggs: 160 each year.

Egg colour: Creamy-white or brown-tinted.

Weight: Hen 2.7 kg (6 lb): Cock 3.6 kg (8 lb).

Temperament: Gentle and inquisitive nature, ideal as free-rangers as well as in a caged run.

The Gold Brahma has richly coloured plumage.

Gold Brahma

A heavy, soft-feathered breed that is tall and stately; its body is deep and its back broad. Its legs are attractively feathered on the outside only, and its tail is rather short. Its head is small, with beetle brows and a triple comb. The Gold Brahma has a richly golden-shaded body, making it stunningly attractive.

Brahmas are an Asiatic breed, originating in the Brahmaputra region of India where they were known as Gray Chittagongs. In 1846, some of these fowls were taken to North America, where they were initially known as Brahma-pootra; this was soon changed and shortened to Brahma. Breed development took place in North America and in 1851 nine Brahmas were presented to Queen Victoria. For many years, only two varieties were known, the Dark and the Light, but later others were developed.

Colour variations: In addition to the Gold Brahma, variations include the White Brahma, Buff Brahma, Dark Brahma, Gold Partridge Brahma, Buff Columbian and Blue Partridge Brahma.

Eggs: 80–100 each year – sometimes more.

Egg colour: Brown and speckled.

Weight: Hen 4 kg (9lb): Cock 4.5 kg (10 lb).

Temperament: Although large, they are calm, friendly and non-aggressive, making them ideal as family pets. They are not easily frightened if handled from when young.

Hamburg

A light and soft-feathered breed with a distinctive elegance that always looks attractive. The variety most often seen is the Silver Spangled, with a white background colour and each feather ending in a round, black spangle. It is mainly kept for its reliable egg-laying nature.

In the late 1500s it was thought that the Hamburg originated in Turkey; Ulisse Aldrovandi (1522–1605), an Italian naturalist, called them Turkish chickens. However, there are stronger claims that the breed began in the north of England more than 300 years ago, where they were known as Moonies. Later, they were called Dutch Everyday Layers and Everlayers. These local names rightly highlight the prolific egg-laying ability of this breed. Holland and Germany have originated several superb varieties.

Colour variations: In addition to the Silver Spangled variety, there are several others, including the White, Black, Silver Pencilled, Golden Spangled and Golden Pencilled.

Eggs: 240 each year.

Egg colour: White, with a glossy shell.

Weight: Hen 1.8 kg (4 lb): Cock 2.3 kg (5 lb).

Temperament: Friendly, active and alert, but often with a skittish and flighty disposition. They are best kept as free-rangers, rather than in a run.

Several Hamburg variations have spangled plumage.

Houdan

In the 19th century this was a very popular heavy breed, for its eggs as well as meat. It was an important meat-providing chicken in France as it matures quickly. Also, in relation to the amount of food it eats it produces a large number of eggs. In addition to its distinctive headdress and beard, its white-speckled and mottled body is handsome.

This is one of the oldest French breeds, originating in Houdan, a town in northern central France. It was recorded in the 17th century but mainly developed in the mid-1800s, when it was known as the Dorking of France because of the fourth toe on each foot. Breeds in its development include Black Polish, Crevecoeur and Dorking, but it is best known for its large headdress and beard that cloak the face. Houdans arrived in England in 1850 and in North America in the late 1850s.

Colour variations: None.

Eggs: 180–200 each year.

Egg colour: White.

Weight: Hen 3.2 kg (7 lb): Cock 3.6 kg (8 lb).

Temperament: Docile and friendly.

Houdans have been known for several centuries.

This Red-mottled Leghorn soon captures attention.

Leghorn

A light, soft-feathered breed with a prolific egg-laying capacity; it is also superb as a table breed. The White Leghorn, developed by British breeders, was taken to North America and for many years played an essential role in their poultry industry. The plumage of Leghorns is soft and silky. Legs are featherless and long, with four toes on each foot. Hens are good layers and do not tend to become broody. In cold weather, the combs of male birds are sometimes damaged; a coating of petroleum jelly provides protection.

Leghorns originated in Italy and were exported from the Port of Leghorn, on the west coast. They were introduced into Britain in the mid-1800s and to North America in 1835, where they were initially known as Italians.

Colour variations: There are many superb varieties and several of them were developed in North America. In addition to the White Leghorn, these include Black-tailed Red, Light Brown, Dark Brown, Black, Cuckoo and Partridge. There are also comb variations within some varieties.

Eggs: About 200 each year – sometimes more.

Egg colour: White.

Weight: Hen 2.5 kg (5 lb): Cock 3.4 kg (7 lb).

Temperament: Often noisy and active, well able to fly over low fences – ensure these are at least 1.8 m (6 ft) high. It is best suited to warm climates and especially when kept as a free-ranger.

Minorca chickens are superb foragers.

Minorca

A distinctive light breed from the Mediterranean region, it was earlier known as the Red-faced Spanish but now popularly as the Minorca. It is the largest and heaviest of the Mediterranean breeds and takes its earlier name from its bright red comb, which contrasts with the large, almond-shaped earlobes. The breed has an upright stance and the most often seen variety is the Black. It is best kept as a free-ranger or in a large, fenced run.

Minorca chickens were imported into the southwest of England in the early 1830s. It is often bred for exhibition purposes but nevertheless still retains its ability to produce plenty of eggs.

It is a breed which has the distinction of being the favourite chicken of the legendary Thomas Edward Lawrence, popularly known as 'Lawrence of Arabia'.

Colour variations: Apart from the Black Minorca, there are White and Blue variations. With some varieties, there are variations in the shapes of their combs.

Eggs: 240 each year.

Egg colour: White.

Weight: Hen 3.2 kg (7 lb): Cock 3.6 kg (8 lb).

Temperament: Its distinctive and independent nature (a characteristic that perhaps particularly appealed to T.E. Lawrence) fosters a liking for the freedom of an open field.

New Hampshire Red

Heavy, soft-feathered, dual-role chicken. Its plumage is fluffy and full, with more orange than that of Rhode Island Red chickens. Males have light yellow-orange hackle feathers. Their legs are yellow and the lower thighs muscular and large.

This all-American breed originated in the New England State of New Hampshire in the early 1900s; it was recognized as a breed in 1935. It has developed from the Rhode Island Red through selective breeding into the handsome and distinctive breed seen today. It is hardy and well able to survive low temperatures, yet still lay plenty of eggs.

Colour variations: None.

Eggs: 200–220 each year (sometimes fewer).

Egg colour: Brown and tinted.

Weight: Hen 2.9 kg (6 lb): Cock 3.8 kg (8 lb).

Temperament: Slow and stately disposition. It is not a high flyer and therefore does not need high fencing. Its placid nature makes it easy to tame as a family pet. They are equally happy as free-rangers as when kept in a run.

New Hampshire Red chickens were developed in North America.

The Old English Pheasant Fowl is a hardy breed.

Old English Pheasant Fowl

A rare light breed and, despite its name, not a pheasant (it gained its name only because of the resemblance). It is usually kept for egg production, although the flesh is 'tasty' and good quality. Hens have a light, rich bay colour with beetle-green tail feathers, while males are rich mahogany-red with green-black tail and feather tips. Their feet and legs are slate-blue. It is a cold-hardy breed and survives well in cool climates. Additionally, it is best kept as a free-ranger. Hens tend to become broody quickly and are ideal mothers.

As a breed it has been known for more than a hundred years and earlier was called the Copper Moss, Golden Pheasant, Old Fashioned Pheasant and Yorkshire Pheasant. It was most popular in northern British counties and has been called the Old English Pheasant Fowl since 1914.

Colour variations: In addition to the above description, there are two varieties – the Gold and the Silver. The gold form has rich bay colouring with darker markings; the silver has white feathering with beetle green-black markings.

Eggs: 160–200 each year – pullets do not start to lay until seven months old.

Egg colour: White or cream – sometimes tinted light brown.

Weight: Hen 2.5 kg (5 lb): Cock 2.9 kg (6 lb).

Temperament: Active and with a habit of roosting in trees. Therefore, give it plenty of space.

Onagadori

This rare, light breed is in the genetic heritage of the Yokohama (see page 52) and so it is not surprising that it has a long tail, often 3 m (10 ft) long, although ones 10 m (35 ft) have been recorded. The ability to develop a long tail results from its tail feathers never moulting, unlike other breeds. It is more than likely that you will never see this breed other than in books or when visiting specialist breeders in Japan. It is raised for ornamental purposes only and requires regular grooming and attention. It is most often seen in its white form, but there are variations. It is only suited to warm and dry climates. This breed is so respected and treasured in Japan that it has been granted preservation status by the Government.

Colour variations: In addition to the White variety, there are the Black Breasted Red, Black Breasted Silver and Black Breasted Golden.

Eggs: About 12 each year.

Egg colour: Light brown.

Weight: Hen 1.8 kg (4 lb): Cock 2.3 kg (5 lb).

Temperament: Dignified, noble and docile.

The Onagadori is a truly memorable breed.

The Plymouth Rock is an all-American breed.

Plymouth Rock

This well-known breed originated in North America and is in the parentage of several breeds, including the popular Barred Rock (see page 33).

The Plymouth Rock is a dual-purpose heavy breed, ideal for egg production as well as eating. They are handsome birds, with long, broad deep bodies.

The first Plymouth Rocks were exhibited in Boston, USA, and claimed to be a cross between several breeds, including Cochin, Brahma, Java, Dominique and Minorca.

Colour variations: In addition to the Barred Rock, there are White and Buff forms.

Eggs: 200 each year.

Egg colour: Light to medium-brown, sometimes with a shade of pink.

Weight: Hen 3.4–3.6 kg (7–8 lb): Cock 4.5 kg (10 lb).

Temperament: Well-mannered, quiet and docile, making it an excellent pet for children. Its egg-laying ability is prolific and not diminished by cold weather.

Poland

Sometimes known as Polish, this light breed is distinguished by its turban-like, rounded and crested head feathers. It is not a breed for the inexperienced chicken enthusiast as the crest needs regular attention to ensure it does not become wet or dirty; this encourages the presence of mites and eye infections. The breed is mainly kept for exhibition, although it does produce about 100 eggs each year. Bearded forms were known in Holland in the 16th century and arrived in England in 1816. The Non-bearded type has different ancestry and is claimed to have originated in a region around the Baltic states, a region then controlled by Poland.

Colour variations: These are extensive and embrace both Bearded and Non-bearded forms in colours including White, Buff, Silver, Golden and Blue. There are also attractively 'laced' forms.

Eggs: 100 each year.

Egg colour: White.

Weight: Hen 2.3 kg (5 lb): Cock 2.9 kg (6 lb).

Temperament: Can be temperamental and requires great care.

This Golden-laced Poland has a majestic appearance.

Members of the Redcap breed have distinctive combs.

Redcap

A light breed ideal for both egg and meat production. Its name is derived from its large and symmetrically shaped rose comb. It has a broad body and well-fleshed breast, which has made it ideal for eating. The wings, which closely fit the body, are red-brown with beetle-green webbing. The feet and legs are lead-coloured.

It is at its best when kept as a forager, and needs a high fence as it is a good flier. Its comb is sometimes damaged by cold weather and may need to be covered in a coating of petroleum jelly.

Native to Britain and sometimes known as the Derbyshire Redcap. Breeds thought to be in its ancestry include the Old English Pheasant Fowl, Golden Spangled Hamburg, Black-breasted Red and Dorking.

Colour variations: None.

Eggs: 180–240 each year.

Egg colour: White.

Weight: Hen 2.5 kg (5 lb): Cock 2.9 kg (6 lb).

Temperament: Hardy and active.

Red Dorking

Heavy, soft-feathered breed with a long pedigree. It is kept for both its egg-laying ability and meat. Incidentally, egg-laying is more prolific in spring and summer than during autumn and winter. The Red Dorking has a distinctively red chest and this, together with its low body and stately walk, makes it a handsome breed.

Like Faverolles, Houdans, Sultans and Silkie breeds, it has five toes on each foot (there are three at the front and two at the rear).

Few breeds of chicken have such a long heritage as the Dorking. Fowls in its ancestry are claimed to have been introduced from Italy to the British Isles by the Romans, about 2000 years ago. It is claimed to be the first epicurean bird in England, with capons raised for banquets.

Colour variations: In addition to the Red Dorking, there are Silver-grey, White and Cuckoo varieties. There are also comb variations.

Eggs: 120–130 each year – sometimes more.

Egg colour: White – sometimes with a pinkish tinge.

Weight: Hen 3.2–4.5 kg (7–10 lb): Cock 4–5.4 kg (9–12 lb).

Temperament: Docile, calm and quiet. It makes an ideal family pet.

The Red Dorking has a beautifully coloured chest.

The Rhode Island Red originated in North America.
Opposite: Cock-birds invariably sound-bite their territory,
especially in the early morning. Noisy but nice.

Rhode Island Red

A well-known medium to heavy, soft-feathered, dual-purpose, North American breed that is superb in itself as well as used to create other breeds. The prolific egg-laying characteristic is passed down through the male line and can be seen in the New Hampshire Red (see page 44).

The Rhode Island Red was developed to be able to withstand cold winters and at the same time lay plenty of eggs. It has dark, rust-coloured feathers, although maroon to nearly black varieties are known. Distinctively, they have red-orange eyes, yellow feet and red-brown beaks.

The development of the Rhode Island Red began in the town of Little Compton, Rhode Island, in the 1800s. The breed was first exhibited in Boston in 1880 and is a mixture of several breeds, including Black-red fowls of Shanghai, Malay and Java. Not surprisingly, it is the official state bird of Rhode Island.

Colour variations: None.
Eggs: 200–230 each year – sometimes more.
Egg colour: Light to dark brown.
Weight: Hen 2.9 kg (6 lb): Cock 3.4–3.8 kg (7–8 lb).
Temperament: Friendly, alert and good-natured.

Scots Dumpy

The ancestry of this light, Scottish chicken is long and legendary. In appearance, the Scots Dumpy resembles its name, dumpy and low, but this belies its superb qualities. It is hardy, able to withstand the rigours of winter weather, strong wind and low temperatures. It is a wonderful mother, good layer and excellent provider of meat for the table.

It tends to waddle and its body is low to the ground; for this reason, it is not suited to life in long, wet grass. There is no set colour for the Dumpy, but Cuckoo, Black and White are the ones mostly seen. The 'Standard' for the breed encompasses any pattern and colour that is allowed for Game Fowl. It has mottled, short legs and four toes on each foot.

Earlier, this breed was known as Creepies, Crawlers, Stumpies, Hoodies, Dadies and Bakies. Archaeological records record it in the British city of York in the 11th century.

Claims are made that the Scots Dumpy's acute hearing enables it to hear approaching marauders wincing when treading on thistles!

Colour variations: In addition to the usual ones (Cuckoo, Black and White), there are Brown, Silver and Gold forms.
Eggs: 120–150 each year.
Egg colour: Creamy-white – some tinted beige and brown.
Weight: Hen 2.7 kg (6 lb): Cock 3.2 kg (7 lb).
Temperament: Usually docile, but males can be aggressive.

The Scots Dumpy is resilient in cold weather.

The Scots Grey often harmonizes with its surroundings.

Scots Grey

Superficially, its coat is similar to the Barred Plymouth Rock and Cuckoo Maran, with evenly striped black barring on a steely-grey background. It has a light but large body (sometimes weighing more than the amounts indicated below), long legs, attractively curved body and a long neck. It is a dual-purpose breed, producing plenty of eggs and having a good amount of flesh. Hens are not usually inclined to become broody.

Originating in Scotland and earlier known as Scotch Grey, it is recorded as a barnyard breed in the 16th century. It is thought to be in the ancestry of Dorking and Malay breeds.

Colour variations: None.

Eggs: 200 each year.

Egg colour: White or light cream.

Weight: Hen 2.3 kg (5 lb): Cock 3.2 kg (7 lb).

Temperament: Independent, active and alert and best kept as foragers, rather than in a run.

Spanish

Correctly known as the White Faced Black Spanish, it is considered to be the oldest of Mediterranean fowls. It is a light breed, with comb and wattles distinctively red but the rest of the face white. The feathering is a lustrous dark greenish-black, with clean, feather-free legs. Beaks are black and eyes dark brown, while the shanks and toes are dark slate.

It is claimed to have been the first breed of chicken introduced into North America: the Spaniards took them to the Caribbean during early colonization. Slightly earlier, they had been bred in Spain and known as The Fowls of Castile and The Fowls of Seville.

Colour variations: In addition to the form described above, there are both Blue and White forms, but these are not officially recognized.

Eggs: 150–200 each year – but can vary widely.

Egg colour: White – and large.

Weight: Hen 2.7 kg (6 lb): Cock 3.2 kg (7 lb).

Temperament: Noisy and active, and while some examples are friendly others have a mean disposition.

The Spanish is easily recognizable by its white face.

Sultans are dainty and distinctive.

Sultan

Distinctive light breed displaying profuse feathering and large, puffy crests, beards, long tails and dramatic foot feathers. It is predominantly white, with pale blue or white legs; each foot has five toes. Most Sultans are kept for their ornamental qualities.

A beautiful and decorative breed, originating in Turkey and particularly popular with the Turkish sultans, they were highly prized and kept in palace gardens as 'moving flowers'. In the Turkish language they were known as *Serai-Tavuk*, which translates as 'fowls of the Sultan'. So prized was the breed that only the Sultan and the ruling class were allowed to keep them. It was not until 1854 that examples of the breed arrived in Britain. It appeared in North America in 1867.

Colour variations: None.

Eggs: 180 each year.

Egg colour: Small and white.

Weight: Hen 2 kg (4.5 lb): Cock 2.7 kg (6 lb).

Temperament: Quiet, docile, friendly and content to be confined in runs. Sometimes it is bullied by larger and more active breeds.

Sumatra

Distinctive light breed with beautiful shiny plumage and a dark, greenish-black face. Males have long tails with cascading feathers that almost touch the ground. Its legs are black to dark; bottoms of feet are yellow. Males often have three or more spurs on each leg.

Native to the island of Sumatra and parts of Malaysia, it is now primarily kept for exhibition. In the mid-1800s it was introduced into Europe and North America for cock fighting.

The breed has also been known as the Sumatran Pheasant, the Java Pheasant Game Bird, the Sumatran Game Bird and the Black Pheasant.

Colour variations: In addition to the black form, there are Creamy-white and Blue varieties.

Eggs: 150 each year.

Egg colour: Creamy-white – sometimes tinted.

Weight: Hen 1.8 kg (4 lb): Cock 2.3 kg (5 lb).

Temperament: Graceful, flighty and keen to fly. Therefore, it needs either an aviary or high fences to contain it.

The Sumatra has an alert and questioning face.

Welsummers have a mixture of breeds in their ancestry.

Welsummer

Light, soft-feathered breed with a full breast and deep chest. Its back is broad and its tail held high. It has a single, bright red comb, strong and short beak and almond-shaped earlobes. Its legs are yellow, but pale during summer and when in strong light. It is equally suited to free-ranging as being in an enclosed run. It tends to lay more eggs in summer than in winter, especially if the climate is cold. It is most often seen in a Partridge form: male birds have a deep, bright orange-red collar and black with brown mottling on the breast and thighs; females have a reddish-brown ground colour, with each feather having small black stippling and a light yellow shaft.

The Welsummer has a number of breeds in its parentage, including the Partridge Wyandotte, Partridge Cochin, Partridge Leghorn, Barnevelders, Rhode Island Reds and Croad Langshan. It originated between 1900 and 1913 in an area alongside the river Ysel, to the north of Deventer, Holland, and wasnamed after the village of Welsum. In 1928 it was introduced into Britain and became prized for its large, brown, mottled eggs.

Colour variations: Apart from the partridge form, there is a Silver Duckwing variety.

Eggs: 200–240 each year.

Egg colour: Large, brown and mottled.

Weight: Hen 2.7 kg (6 lb): Cock 3.2 kg (7 lb).

Temperament: Alert, active, friendly and easy to handle. The hens are not good mothers, and do not become broody until late spring.

Yokohama

Its long tail makes this breed an ideal bird for exhibition. It is a light breed with feathers often cascading from its tail and dragging along the ground. For this reason it needs to be kept in a large run with a base that can always be kept clean and dry.

There are several colour variations (see below), with a single, pea or walnut comb.

As its name implies, the Yokohama is a Japanese breed. It did not originate in the port of that name, but it was from there that French missionaries first exported the breed to Europe. The breed's development is complex and owes its current appearance to two different Japanese Cultural Monument breeds, the Long-tailed Onagadori and the Minohiki ('Saddle Dragger').

Colour variations: These include the White, Red Saddled, Black-red and Duckwing.

Eggs: 150 each year.

Egg colour: Creamy or cream-tinted.

Weight: Hen 1.8 kg (4 lb): Cock 2.3 kg (5 lb).

Temperament: Dignified and slow.

The Yokohama is a chicken connoisseur's breed.

Hybrid chickens

Interest in the development of hybrid chickens began in the late 1940s when commercial egg producers began to respond to an increasing demand for eggs. There was a desire for breeds that would lay 300-plus eggs each year. In addition, such breeds should show a good conversion rate of food eaten to the production of eggs.

Hybrid chickens are the progeny of parents from two different and distinct pure-bred lines. This results in hybrid vigour in their offspring. Shuffling and rearranging the genetic characteristics of two pure breeds to produce another breed has both desired and undesired results – but usually they offer undeniable advantages.

Advantages of hybrids
- Number of eggs produced is high.
- Reliably produce eggs throughout the year.
- Usually easy to look after, rarely becoming broody.
- Food-to-egg conversion rate is high.
- Usually available from suppliers throughout the year.

Disadvantages of hybrids
- Usually more expensive to buy than a pure breed, but this has to be balanced against the opportunity to have increased numbers of eggs or a more meaty carcass
- Hybrids do not reliably pass on their characteristics to their progeny (indeed, some hybrids are sterile). Therefore, if you have a few hybrid hens, and they are mated with a male bird, do not expect their progeny either to resemble them or to have their characteristics.
- After their first season of laying, the number of eggs usually decreases. However, this characteristic is also naturally present in other breeds.
- When in a group of pure-bred chickens, hybrids sometimes have a tendency to become aggressive.
- Hybrids are often prone to egg-laying problems.

- It is essential to buy hybrids from a reputable supplier who can assure you that the chicks or point-of-lay pullets are what you are expecting.
- They do not usually make good family pets.

Pure breeds
These are breeds such as Rhode Island Red, Marans and Leghorns which are well established and have been popular with chicken enthusiasts for many years. When used for breeding – and using males and females of the same breed – they produce true replicas of themselves.

Each pure breed exhibits the personality, egg-laying and meat-producing characteristics of its breed. This is unlike hybrids which show a combination of two distinct breeds.

Pure breeds usually lay eggs for more seasons than hybrid types, but are more likely to become broody and, consequently, stop laying for a time. Additionally, some pure-bred chickens are difficult to source and can be expensive to buy. Nevertheless, for the majority of home poultry keepers pure-bred chickens are the first breeds to consider when starting to keep poultry.

There are many pure breeds suitable for home poultry keepers (see pages 31–52 for a range of light and heavy breeds); also pages 56–61 for 'miniatures' and 'True Bantams').

RANGE OF BANTAM BREEDS

Bantam breeds

Bantam breeds of chickens are popular and especially suited where space is limited. They are smaller than large-fowl breeds (see pages 31–52) and therefore less expensive to keep. There are two main forms of these small, inquisitive, colourful and amusing chickens: some are smaller versions of large breeds and known as 'Miniatures', while others are 'True Bantams' and not related to larger breeds.

Usually, the definition of a Miniature is that it is about a quarter of the size of its corresponding large breed. Nevertheless, it must have the same characteristics and usually the same colours as its larger counterpart.

The sizes of eggs laid by Miniatures and True Bantams are less than those from most large-fowl breeds. Eggs from large fowls weigh 50–75 g (2–3 oz), whereas those from Miniatures and True Bantams are usually 40 g (1¹/₂ oz).

Large-fowl breeds with miniature versions

★ Araucana
★ Australorp
★ Barnevelder
★ Brahma
★ Cochin
★ Croad Langshan
★ Dorking
★ Faverolles
★ Frizzle
★ Hamburg
★ Leghorn
★ Maran
★ Minorca
★ New Hampshire Red
★ Orpington
★ Plymouth Rock
★ Rhode Island Red
★ Silkie
★ Sussex
★ Welsummer
★ Wyandotte

Confusing opinions

The determination of a Miniature or True Bantam can be confused by breeds being classified differently from one country to another.

Silkies are ideal family pets. They are not True Bantams, but small versions of large-fowl breeds.

Dutch Bantams are thought to be the smallest True Bantam breed.

A–Z of True Bantam breeds

True Bantams, unlike Miniatures, do not have large counterparts. Many are kept as pets or for their decorative qualities and therefore their limited egg production is not a problem. Here is a selection of the most popular True Bantam breeds.

Belgian

This breed has a rather dumpy body but nevertheless has a pleasing appearance, with a low wing carriage, short back and high tail.

It has a rather complex background and with three different forms. These are the d'Uccle (single comb, bearded, whiskers and feathered legs), the d'Anvers (bearded, rose comb, whiskers and clean legs), and the d'Watermael (similar to the d'Anvers, but with three leaders from the back of its comb). Additionally, all three of these forms are seen in several colour variations; their weights also vary.

Colour variations: There are three (see above).
Eggs: Tiny and very few.
Egg colour: Creamy-white.
Weight: Hen 620 g (22 oz): Cock 740 g (26 oz).
Temperament: Friendly and lively, with a jaunty nature.

The Belgian d'Watermael has a lively temperament.

Dutch Bantams are smart, jaunty and lively.

Dutch

Sometimes known as De Hollandse Krielan and, originally, Old Dutch, this is one of the smallest True Bantam breeds. It has a short back, full and high breast, and an attractive upward stance. The long and large wings are carried close to its body. The single red comb is attractive, with five serrations. The earlobes are white, the beak is short and the legs are slate-blue.

Colour variations: Silver Partridge, Gold Partridge, Yellow Partridge, Blue-yellow Partridge, Blue-silver Partridge, Blue Partridge, Red-shouldered White, Cuckoo Partridge, Cuckoo, White, Black, Blue and Lavender.
Eggs: Tiny, and very few.
Egg colour: Light tint.
Weight: Hen 400–450 g (14–16 oz): Cock 480–540 g (17–19 oz).
Temperament: The hens make good mothers, although being small they are unable to cover a large number of eggs at one time. Egg laying is often limited to the summer months.

Japanese

Also known as Chabo, the Japanese True Bantam breed has a fascinating and highly distinctive shape. Its tall and upright tail, short back and short legs create a breed with a wading gait. Its body appears to tilt forward as it walks. The large, single comb is evenly serrated.

The tail in male birds tends to come forward, almost touching the bird's neck. The hen's tail is more upright. There are several colours (see below).

It is thought to have originated in Southeast Asia – Malaysia, Java and Indonesia – but was developed in Japan. In the mid-1600s it appeared in Japanese and Dutch art. The name Chabo is derived from Chabol, meaning dwarf.

Colour variations: The most popular ones are Black-tailed White, Black and White, and Birchen, but there are others, including Black Breasted Red, Black, Blue, Gray, and Black-tailed Buff.

Eggs: Tiny – and very few.

Egg colour: Creamy-white.

Weight: Hen 620 g (22 oz): Cock 740 g (26 oz).

Temperament: Friendly and ideal as a pet but male birds can, occasionally, be aggressive. They do not peck deeply into lawns and can be allowed to wander over mown grass.

© STUART SUTTON

Nankin Bantams have an upright carriage.

Nankin

This is a delightful True Bantam, with a dainty nature and appearance. It has a high, arched neck and a tail carried upright and with feathers spread out. It is a hardy breed and slow to mature. The beak is strong and slightly curved downwards. It is only seen in a buff colour, but there are two comb forms – single or rose.

It is thought to have originated in India and been used in the development of the Sebright Bantam (see page 61). Additionally, it has been used to create buff-coloured forms in other breeds.

Colour variations: None.

Eggs: Tiny – and very few.

Egg colour: Creamy-white.

Weight: Hen 620 g (22 oz): Cock 680 g (24 oz).

Temperament: Not always friendly.

© DI WOOLFE

Black-tailed White Japanese Bantams are very popular.

Pekin Bantams have 'tea-cosy' outlines.

Pekin

A popular and widely kept True Bantam, only 20–30 cm (8–12 in) high even when holding its head upright. Its carriage is distinctive, with a forward tilt that in females sometimes results in its breast almost touching the ground. Its body is round, giving the appearance of a 'walking tea-cosy', with feathers splaying out, especially in males. It has a single comb and feathered legs, again mainly in males. There are many claims about its ancestry, including examples being stolen from the Emperor of China in about 1860. Another claim is that examples were given to Queen Victoria in the 1830s and that these were crossed with other breeds.

Colour variations: There are many differently coloured forms, including Black, Black Mottled, Blue, Buff, Colombian, Cuckoo, Gold Partridge, Lavender, Partridge, Red, Silver Partridge, and White. There is great interest in these bantams in North America.

Eggs: Small.

Egg colour: White to tinted.

Weight: Hen 570–680 g (20–24 oz): Cock 680–790 g (24–28 oz).

Temperament: Friendly, docile and gentle, all characteristics that make them ideal as family pets, especially for children. Hens have a tendency to become broody and therefore make good mothers.

Rosecomb

Occasionally known as the Rose Comb Bantam, this True Bantam has a similar outline to the Dutch True Bantam (see page 58). The rose comb is square at the front and ending in a long, pointed and slightly upward spike at the rear.

There are several varieties, all with different colours: both the Black and the Blue varieties have slate-coloured toes and shanks. The White variety has pinkish-white toes and shanks. It is a breed with distinctive white, large and round earlobes.

This is an old bantam variety that is thought to have originated in Britain, with Black Hamburgs and other bantams in its blood.

Colour variations: As above.

Eggs: Tiny – and very few.

Egg colour: White to cream.

Weight: Hen 510 g (18 oz): Cock 620 g (22 oz).

Temperament: Usually friendly, although cocks can be aggressive. They are flighty and active and thrive in cool as well as moderately warm climates.

Rosecomb Bantams have upright tail feathers.

Sebright Bantams possess distinctive and unforgettable plumage.

Sebright

This is perhaps the most unusually coloured of all True Bantam breeds. There are two differently coloured varieties, the Silver form, which appears more white than silver, and the Gold type, with a rich and deep colour. Both of these have feathers distinctively edged in black, creating a dramatic and memorable feature.

This British breed was developed by Sir John Sanders Sebright in the early 1800s. It originates from a cross between a bantam and a Poland (see page 46).

Colour variations: As described above.

Eggs: Tiny – and very few.

Egg colour: White or creamy-white.

Weight: Hen 570 g (20 oz): Cock 620 g (22 oz).

Temperament: Active, alert and long-lived. It likes to roost in trees and is a difficult bird to keep for novice poultry keepers.

WAYS TO KEEP CHICKENS

Housing chickens

Methods of keeping chickens range widely; some are traditional, with chickens wandering freely, others evolved after the mid-1900s when the demand for food dramatically increased. A few of these ways are well suited for use by home poultry keepers and include a poultry shed with a scratching run. Others are commercial and, to many people, show little concern for the welfare of chickens.

Free-ranging chickens characterize the relaxed nature of the 'good life'.

Free-range system

This is the most natural way to keep chickens if you have sufficient space – but for most home poultry keepers it is just not possible.

- **It is an idealistic method**, where chickens roam free but within a fenced area. A recently harvested field with stubble still standing is a superb venue, where chickens scavenge for food, picking up seeds as well as ridding the land of insect pests. Additionally, their droppings help to increase the land's fertility.

- **Secure fencing is essential** to keep out predators, especially foxes who may not even be deterred by a fence 1.5–1.8 m (5–6 ft) high. Additionally, bases of fences need to be dug into the soil (see page 78 for details of vermin-proof fencing).

- **A hen house is essential** to provide warmth and protection at night. One on wheels enables movement from one area of a field to another, so that the total area is evenly grazed by your chickens.

Alternatively, an ark is easily moved. Where only a few hens are kept in an ark, they are especially at risk from low winter temperatures. However, when there is a large number present their bodies collectively raise the temperature. Conversely, in summer the low cubic air capacity in an ark may rise dramatically and cause difficulties for your hens.

- **Chickens in fields during summer** need shade (see 'Selecting a suitable site', page 72).

- **Food and water need to be dispensed** to the hens every day and therefore you should ensure that a supply of clean water is readily available. If the hens are outside during winter and water is available in the field through a standpipe, in autumn or early winter cover it with insulative materials and a plastic wrapping.

Fold method

This is where chickens are kept in a poultry house with an attached wire-netting enclosure. It is a safe way to keep hens as they are completely protected yet have the ability to wander and scratch.

There are many variations on the fold system; some have the living quarters on wheels, with the run part attached and secure against vermin.

This visual update on a scarecrow initially deters vermin and adds amusement to keeping chickens.

Semi-intensive system

This gives chickens a good life as they are able to wander in a field during daytime and be safely locked up at night. But the system does need a farm with a couple of spare fields, in which the chickens can graze during the day, as well as a barn in which they can be kept at night.

The floor of the barn needs to be covered in straw. If the weather is wet or extremely cold, the hens can be kept indoors during hours of daylight – as well as at night. Throughout summer, chickens can graze in a field and pick up and eat seeds and insects from the land.

To prevent a field becoming excessively grazed or contaminated by pests and diseases, regularly alternate the two fields as daytime homes for your chickens.

Deep-litter system

Similarly to the semi-intensive system, this method also needs farm-like facilities and therefore is not a possibility for home poultry enthusiasts. Hens are homed in a barn or very large shed, without being allowed out of doors at all.

The birds are kept on layers of litter or straw regularly topped up and completely changed annually. Although totally confined indoors, hens are able to wander freely. The system provides warmth and security for them.

A variation on this system is to allow chickens to wander outdoors during daytime in a wire-netting enclosure with a thin layer of straw on the ground. This variation is sometimes known as the straw-yard system.

Battery system

Previously, this was an accepted commercial method of keeping hens in the UK and Europe, as it still is in many other parts of the world, but the barren battery cage is now illegal in this country and across the EU, thanks to welfare concerns for the hens and pressure from consumers.

Battery systems do not generate happy chickens!

Conventional 'barren battery cages' were banned under EU legislation on January 1st 2012. For 13 years prior to that date, all EU producers were advised to replace barren battery systems with 'free-range', 'barn', or 'enriched cage systems', which provide more space and improved conditions for the hens.

The illegal barren cages allowed 550 sq cm for each bird, while the enriched cages now give each hen 600 sq cm. In addition to extra space, the new cages must also contain a scratching area, nesting box and perch, which allows the hens to practice more of their instinctive behaviour.

The UK's only registered hen charity, the British Hen Welfare Trust, has stated that it would rather see all laying hens in small flocks with access outdoors to free range, but whilst there remains consumer demand for cheap caged eggs, it would rather those caged eggs were produced in the UK where welfare standards meet EU regulations.

Egg producers in the UK have met the requirements of the new legislation, while around one-third of eggs produced throughout the rest of Europe are still from illegal barren battery cages, which, when viewed with the fact that more than half of all eggs produced in the UK are from free-range and barn systems, provide additional reasons to check the label and try and buy British eggs. It is relatively easy to 'hide' battery cage eggs in processed foods such as cakes, pasta, ice cream etc, and if the label does not say British or free range, it may have been produced by hens living in barren battery cages overseas.

Bustling and clucking free-ranging chickens soon become aware of feeding times.

Poultry sheds and equipment

The range of sheds and pens in which chickens can be kept is amazingly wide. Some are legacies from earlier years and, at first sight, appear antiquated and dilapidated – but are usually functional and provide weatherproof homes for chickens.

Sometimes, creating a secure home for chickens is a matter of adaptation, perhaps buying the living quarters and building a scratching pen from materials recycled from earlier poultry houses. With generous help and advice from poultry-keeping friends – as well as a hammer, saw and few nails – many sound and long-lasting poultry sheds have been constructed on low financial budgets. Wire-netting and wood is often available at a modest cost from reclamation depots. And this is good for the environment as well as for you.

Newly constructed poultry sheds are available and a range of them is described within this chapter.

Ark

Distinctive design and basically triangular from its end view. About one-third of the floor space forms an enclosed sleeping area, with the other two-thirds a place for walking and scratching. The sleeping area has a pop-hole linking it to the scratching region, with a shutter that can be raised during the day and closed at night. Sometimes, nesting boxes are integral with the end of the sleeping area.

The sizes of arks vary, with the largest 6 m (20 ft) long and 1.8 m (6 ft) wide and able to offer a home to 20 or more hens. For a home garden, however, an ark half this size is sufficient for five or six hens.

An ark is ideal for free-range chickens. It also has the advantage that it can be moved, by two people, to a different piece of ground.

As well as suiting free-range chickens, it is sufficiently small and portable for hens kept in a back garden or yard.

Because of its small nature and limited insulation, the air inside the sleeping quarters may rise dramatically in summer and fall in winter.

Walk-in shed

A traditional design, similar to a large garden shed and usually with a lean-to shaped roof. It is static and large enough to enter with ease. Usually, there is a row of nesting boxes along one side, with access to them from the outside so that eggs can be collected without entering the shed.

A pop-hole, with a shutter that can be raised during the day and closed at night, is usually positioned at the front of the shed, with a ramp so that chickens can easily reach the ground.

Secure door fastenings are essential. Additionally, provide ventilation through a wire-meshed window opening, with a hinged shutter that can close over it to prevent vermin gaining access and wind and rain blowing in.

During the daytime, when your chickens are outside the shed, they will need corralling to prevent them wandering off; a wire-netting fence 1.2–1.5 m (4–5 ft) high is usually sufficient to keep out daylight marauders.

Both wheeled and static hen houses provide safe havens for chickens – especially at night.

Hen house on wheels

As its name explains, this is a chicken house on wheels. It is ideal where free-ranging hens are kept in a field and usually accommodates more chickens than an ark. It often needs two or more people to move it – or one person and a tow from a robust, four-wheel-drive vehicle.

It is usually of sturdy construction (movable sheds have to be well-made to withstand movement over rough and uneven ground). The wide-rimmed metal wheels are small but large enough to keep the shed's base about 10 cm (4 in) above the ground. Similarly to a walk-in shed, a secure door with good fastenings is essential, as well as a wire-netting window fitted with an external shutter.

When used in a field, a pop-hole, with a shutter able to be raised during the day and closed at night, is essential. Make sure an easily accessible ramp enables chickens to reach the ground safely.

A row of nesting boxes along one side allows eggs to be collected without having to enter the house.

Whatever the poultry shed's structure, sloped ramps are essential to enable easy access for chickens.

There are no strict rules about using a hen house on wheels and much depends on its age and condition; when new it can be easily moved but with wear invariably becomes rickety. At this stage, the answer is to use it statically with a large scratching run added to one side.

Poultry shed with a scratching run

This is a static arrangement and ideal for use where foxes and other vermin are prevalent. Basically, it is similar to a walk-in shed but with a large, wire-netting scratching pen added to one side. Ensure the roof is covered in strong, well-secured

wire-netting. Foxes are adept at leaping on to a wire-framed roof and pulling the netting apart.

Nesting boxes, accessible from the outside, are essential, as well as a pop-hole with a shutter that can be raised during the day and closed at night. Make sure an easily accessible ramp enables the chickens to reach the ground.

Ventilation within a poultry shed is essential and this is best achieved with a wire-netting window with a hinged or sliding shutter.

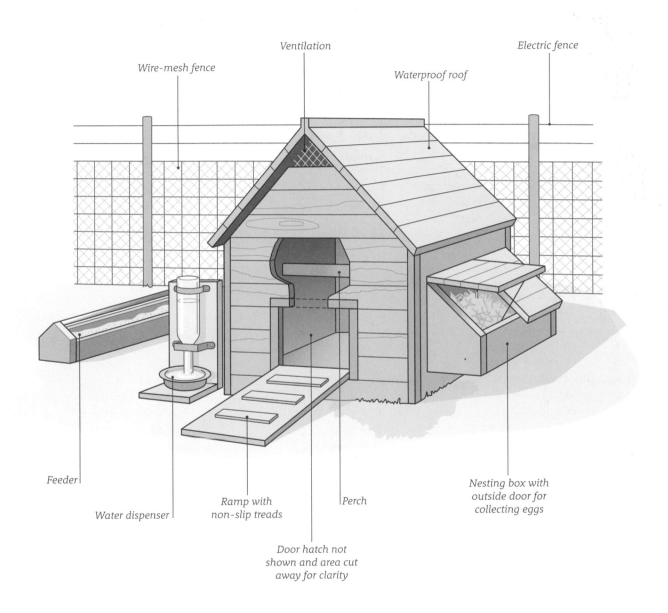

Wire-mesh fence

Ventilation

Waterproof roof

Electric fence

Feeder

Water dispenser

Ramp with
non-slip treads

Door hatch not
shown and area cut
away for clarity

Perch

Nesting box with
outside door for
collecting eggs

A warm, dry and draught-free home is essential, as well as protection from vermin.

The space you will need

It is essential not to squeeze an excessive number of chickens into a hen house, yet to utilize the space effectively. Putting a large number of chickens into a cramped hen house will result in squabbles and the hens pecking at each other.

• **Allow space** for six hens; this will provide a family with eggs throughout the year. Only three hens are needed for a family of two.

• **For their sleeping quarters**, allow a minimum of 1 cubic foot for each average-sized bird.

• **For a covered run** and occasional garden access, each bird needs 3 square feet.

• **Large hens need** about one-third more space than an average-sized chicken.

• **Bantams are able to live** happily in one-third less space than an average-sized chicken.

Selecting a suitable site

Sometimes there is little choice when positioning a hen house in a small garden; invariably it is put out of sight of the dwelling house and in a secluded corner. When there are choices of position, however, here are a few considerations:

• **The movement of air** around a hen house helps to keep chickens healthy and free from disease. However, there must not be strong and direct draughts, especially during winter.

• **A convenient supply of clean**, piped water is needed. Check that tap washers are not deteriorating, allowing water to drip over the surrounding ground. If this happens the area becomes a quagmire in summer and an ice-rink in winter.

Ensure that chickens are not crowded together as this encourages health problems.

Orientating and positioning a hen house

Usually, there are several ways to position and orientate a poultry shed:

★ Avoid the bottom of a slope. This is where cold air migrates and becomes stagnant. Air at the base of a slope is much colder in winter than that which is further up the gradient.

★ If a poultry house is positioned on the side of a slope, ensure that storm water cannot run into the enclosure.

★ The house should face the sun to take advantage of all possible sunlight; this will enable it to remain warm and dry.

★ The closed or rear side of a chicken house should be positioned to give protection from cold, prevailing wind.

★ Where the main winter weather is coming from the west, the poultry house should face east. Conversely, where the main winter weather is coming from the east, position the chicken house so that it faces west.

• **Good soil drainage** reduces the risk of diseases and pests.

• **Choose a position where smells** created by chickens will not waft over neighbouring properties.

• **Essentially, the site and housing** should provide your chickens with a dry place in which to live, as well as fresh air and no draughts.

Chickens are adaptable to their environment and delight in scratching and investigating the ground.

Chicken house fixtures and equipment

Within a chicken house – as well as immediately outside and perhaps within a scratching area – several fixtures and pieces of equipment are essential for the well-being of your chickens. Here is a range of them.

Roof

A waterproof roof is vital. Several constructional solutions are available:

• **Tongue-and-groove boarding** covered with roofing felt is ideal. Use thick and strong felt as thin types may only last for a few years, especially when in strong and hot sunlight. Rainwater dripping off trees also causes roofing felt to deteriorate rapidly.

Where one layer of felt overlaps another, use a sealant to prevent water penetrating between them. Ensure that felt along the ends and sides of the roof is securely fastened as this is where deterioration first occurs, especially in exposed and windy areas. Coating the entire roof in a sealant every three to five years ensures a longer life for the felt.

• **Corrugated fibre-glass roofing panels** are a long-term solution to keeping out rain. They need to be overlapped and firmly secured to a strong framework.

• **In earlier years, corrugated-iron panels** were used, but these encourage high temperatures within the poultry house during summer, and cold ones in winter.

Ventilation

To ensure the well-being of chickens throughout the year, a good circulation of air is essential. Stuffy conditions, as well as cold draughts of air, encourage the presence of respiratory problems.

★ **There are several ways** to provide ventilation and one is to have a louvred opening near the top of the shed (perhaps in the apex of a ridge-type roof) and another lower down, covered in wire-netting and with a hinged or sliding shutter for closing at night.

★ **Wherever possible, position ventilation** openings on the side of the shed that is away from strong, prevailing wind.

Small hen houses are easier to move to fresh ground when they have wooden bearers as shown here.

Flooring

The nature of a hen house's floor is influenced by its position. An ark, for example, is just positioned on the ground, while walk-in sheds usually have a firm floor that creates a dry base.

● **Solid concrete creates** a permanent base and prevents rats burrowing in from outside. Preferably, such a base should be slightly raised above the surrounding ground to prevent flood water seeping in.

● **Rot-resistant, tongue-and-groove wood** forms a level and functional floor that keeps chickens warm and dry. However, it has a limited life and needs regular checking, especially when it is getting old.

● **Well-rammed gravel and rubble** produces an inexpensive floor, but may become invaded by rats.

A hen needs privacy when laying eggs.

Nesting boxes

These are essential for hens to lay their eggs in; there should be one nesting box for every three hens. If there are too few boxes it can result in hens squabbling and laying their eggs elsewhere and where they are difficult to find, especially when hens are free-ranging.

● **They are best positioned** so that a hen is able to enter from inside a chicken house, while the egg collector can raise a lid on the outside and lift out the eggs directly into a basket or other container.

● **Where there is a risk of vermin** gaining entry to the inside of a poultry house through a nesting box, fit a simple clasp-like securing device to its outside.

Perches

These are essential to enable birds to roost at night. A hen's foot, like those of all perching birds, has toes that, once gripped on a perch, lock so that she does not fall off while asleep.

● **Never overcrowd birds** on perches – allow at least 20 cm (8 in) of space for each bird. If hens are extra large, increase this to 25 cm (10 in).

● **Use 5–7.5 cm (2–3 in) thick**, clean timber for perches – do not economize and choose a smaller diameter. Make sure they are smooth and do not have splintered edges.

● **Round off the edges of the perches** to make them easier for birds to wrap their feet around.

● **Position perches 60–75 cm (2–2¹/₂ ft) above the floor. No perch should be higher than this, because if a heavy hen jumps from a high perch onto the floor it might become damaged. Bumble foot (see page 176) is sometimes caused by a bird jumping from a high perch.

● **Where possible, make the perches removable** so they can be inspected outdoors in good light for parasitic pests such as Red Mites (see page 166).

● **Perches must be level**; if sloped, your hens will feel uncomfortable and may become disorientated.

● **Where possible, position a wide board** under the perch to catch their droppings.

Feeding and drinking equipment

Clean drinking and feeding equipment is essential and all feeders and drinkers need to be cleaned each day. This especially applies to water dispensers; when in the open they may become contaminated by faeces from wild birds (see pages 172–173 for details of diseases spread by birds).

Hens need access to water at all times and to food several times each day. Where possible, use self-dispensing equipment as if for any reason you are unable to attend to your chickens they will still receive food and water.

The range of equipment is wide, some bought specially for the purpose and others home-made. Here is a range of feeders and, where applicable, the types of food suited to them.

• **Dry mash, corn and grit** can be given through troughs or self-feeding devices.

• **Indoors, wet mash** is best given in a plain, open trough. However, if outdoors – and unless the run is covered by wire-netting – wild birds soon become aware of it and steal the food. There is also the risk of them defecating into them and spreading diseases.

• **Suspended self-feeding devices** are ideal in a large chicken house and where they can be safely positioned.

• **Low drinking devices** are available for chicks – ensure that they cannot fall in!

• **A practical and useful type of feeder for chicks** is one which allows small heads to get at the food, but prevents large birds from taking the food.

Food storage hoppers and containers

The storage of dry mash, corn and grit is best undertaken in hoppers or large, lidded containers. These must be kept in a dry, vermin-proof shed. Preferably, they need to be positioned close to the chickens, but this is not always possible.

Dust baths

To some poultry keepers, these often appear unpleasant activities regularly pursued by chickens, but to chickens they are like visits to a spa town – such joy!

• **A hen's natural behaviour** is take a dust bath to remove parasites such as lice (see pages 166–168 for details of external parasites).

• **A hen will seek an area of fine**, dry earth and, literally, bathe in it, allowing dust to trickle through her feathers and on to her skin, where it discourages parasites.

• **Free-range hens**, as well as those with an outdoor pecking area, readily find a dust-bath area.

• **For dust bath facilities** in a poultry house, a shallow, wooden box filled with fine, silver sand is ideal, but ensure that it does not become wet.

• **Wood shavings** are an alternative dust-bath material but they tend to become scattered by chickens and blown about by strong winds.

Hens like nothing better than a dust bath.
Opposite: Feeding troughs must be kept clean.

Vermin prevention

It is often claimed that wherever you live there is a rat within 4.5 m (15 ft) – or less – of you. Even if you do not see one, be assured that they are able to gnaw through wood and metal to get at your chickens. Foxes are present in the countryside and towns and will soon be aware of your chickens.

Here are a few ways to discourage vermin:

- **Do not leave scraps** of food about.

- **Keep your poultry house clean**, so that smells do not immediately attract them.

- **Wash feeders and watering** devices each day.

- **Store food in vermin-proof** containers in a secure, locked shed.

- **Keep all poultry house** doors and windows shut and locked at night.

- **Regularly check that rats** are not burrowing under poultry house floors and scratching runs.

- **Erect strong boundary fences** as defence against foxes.

Fox-proof fencing

Even if you install top-quality fencing many feet high, do not be surprised if a fox eventually gains access. They are adaptive scavengers and, depending on how hungry they are, will usually attempt entry.

However, here is one way to keep them at bay – at least for a time!

Foxes are forever looking for ways to reach your chickens.

- **Use strong, 2.4–3 m (8–10 ft) long, timber posts** treated with a wood preservative. You may be able to buy posts that have been pressure-treated with a preservative.

- **Hammer these posts 45–60 cm (1–2 ft)** vertically into the ground every 1.8–3 m (6–10 ft) along the perimeter.

- **Use either a sledge hammer** or, preferably, a post-driver to drive each post into the ground. A post-driver is a strong metal tube, sealed at one end, with handles on opposite sides. The open end is slotted over the top of the post and then repeatedly rammed downwards to force it into the ground. It can be used by one person, but preferably two. A post-driver

Strongly linked wire-netting is essential to keep out foxes.

is less likely than a sledge hammer to splinter and damage the top of the post.

• **Secure 18-gauge, galvanized**, 5 cm (2 in) mesh wire-fencing to the posts, with 45–60 cm (1–2 ft) buried in the ground and 1.5–1.8 m (5–6 ft) above soil level.

• **Ensure that the door** is equally secure and has a good, strong base.

Keeping foxes at bay

Country folklore abounds with advice about keeping chickens safe from foxes. One method is to urinate around the perimeter of a hen house. Furthermore, urine from a male is said to be a greater deterrent than that from a female.

DAY-TO-DAY CARE OF CHICKENS

Caring for chickens

Keeping chickens in your garden or back yard becomes a way of life and one packed with amusement from your dedicated team of egg layers. It is also a hobby that demands your attention every day throughout the year. The birds are totally reliant on you for food and water, as well as protection from vermin. They must at no time be neglected.

Eating habits

A chicken's eating habits are dictated by its digestive system, which is radically different from that of mammals. When a chicken eats food, it is rapidly transferred to its gullet, also known as the oesophagus, which leads from the bird's mouth to its stomach. The gullet is a narrow, muscular tube but well able to expand to accept large pieces of food.

The gullet widens into the crop, where food is lubricated to help ease its passage downwards. Sometimes, food is retained in the crop for several hours, where it is increasingly softened. Eventually it reaches the glandular stomach and the gizzard.

The gizzard is the equivalent of the stomach in mammals. If the gizzard is empty, food usually passes directly into it, rather than lingering in the crop.

Feeding requirements

Similarly to humans, chickens need (in addition to clean water) a balanced diet. This is essential to enable a hen to lay eggs over a long period.

The essential parts of a chicken's diet change throughout its life. For example, chicks need more protein than older birds especially when they are in their early weeks.

Here is a range of nutrients, their sources and functions in keeping chicks and chickens healthy.

Proteins
• Function: Body building and repairs to damaged parts.
• Main sources: Fish and blood meal, skimmed milk and soya.

Carbohydrates
• Function: Energy.
• Main sources: All cereal grains.

Fats and oils
• Function: Energy.
• Main sources: Meat and bone meal, fish meal and nuts (groundnuts).

Vitamin A
• Function: Disease resistance and balanced growth.
• Main sources: Grass and maize meal.

Vitamin B complex
• Function: Optimum growth development. Also essential for feather and skin health, aiding reproductive, general health and the functioning of the nervous system.
• Main sources: Cereals, fish meal, yeast and skimmed milk.

Vitamin D
• Function: Strong egg shell formation and healthy growth

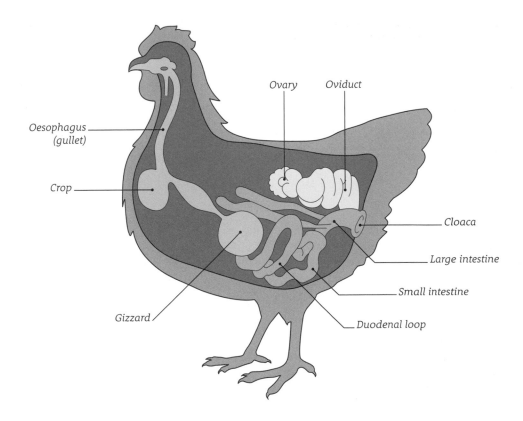

Ovary

Oviduct

Oesophagus
(gullet)

Crop

Cloaca

Large intestine

Small intestine

Gizzard

Duodenal loop

The internal parts of a chicken are remarkably simple – but thoroughly practical.

(including prevention of rickets).
- Main sources: Fish meal, cod liver oil and, essentially, sunlight.

Vitamin K
- Function: Healthy blood.
- Main sources: Grass.

Calcium and phosphorus
- Function: Strong shells and healthy bones.
- Main sources: Bone meal, fish meal and meat.

Zinc
- Function: Feather development and healthy skin.
- Main sources: Usually given in a supplement.

Manganese
- Function: Strong egg shells.
- Main sources: Usually given in a supplement.

Iodine
- Function: Influence and control of metabolism.
- Main sources: Usually given in a supplement, and in seaweed extract.

A question of digestion

A chicken's method of digesting food is simple yet effective, but remains a puzzle to many people. Here are a few popular questions on the subject:

• **Do chickens have a sense of taste?** Yes. Chickens have glandular taste buds associated with the ducts of their salivary glands. These are not highly sophisticated, but sufficient to enable a chicken to determine the food it prefers to eat.

• **What happens when a chicken swallows food?** It passes directly into the gullet and down into the crop; later, it passes into the gizzard (see 'Eating habits', page 82).

• **Can chickens chew?** No. They do not have teeth. However, their hard, horny beaks are tough and, if necessary, well able to break up small pieces of food before it is swallowed.

• **Is grit essential for a chicken's digestive system?** Yes. Because chickens do not have teeth, grit in the gizzard performs the same role. The food-mincing muscular action of the gizzard is made more efficient when grit is present. Therefore, providing a diet that includes grit is essential.

Grass cuttings, kitchen scraps and home-grown vegetables

These 'green' foods are essential for the health of chickens; they not only provide a supplementary diet but help to create interest for hens. In small runs, birds which become bored may interest themselves in unpleasant habits including feather and vent pecking, as well as eating eggs (see pages 176–179 for details of these problems).

While chickens find 'green foods' appetizing and interesting, they are not high in energy and should not be considered as the main part of a diet. Large vegetables, such as cabbages, are best suspended in the run. Small ones can either be chopped up or added to a mash diet (see page 87).

Chickens delight in pecking at food, sorting out parts they particularly like and leaving others for later.

• **Grass cuttings**: After mowing a lawn (it must not have been treated with a weedkiller) collect the cuttings. They can be given fresh and green to chickens, or spread on a clean surface to become dry for collecting, storing and later adding as a winter supplement. However, be aware that chickens which have eaten grass cuttings produce eggs with deep yellow yolks.

● **Kitchen scraps**: This is an easy and inexpensive way to feed your chickens. Potatoes and other kitchen scraps when thoroughly cooked are an ideal food; in this state, they are much more easily digested by chickens than when given to them raw.

Do not use scraps of food containing poultry derivatives (they may be infected with diseases). Also, do not accept gifts of cooked kitchen scraps from neighbours and friends as you will not know their content.

Home-grown vegetables

These can be collected and given fresh to your chickens. Always check that they have not recently been sprayed to control pests and diseases.

The range of home-grown vegetables is wide and includes beans (dwarf, French and runner types), cabbages and other brassicas, carrots, lettuces, parsley, potatoes (do not use when green or sprouting), turnips and swedes.

Chickens are scavengers for food and, if not corralled, completely wander over gardens making a mess of lawns and beds.

Wild plants

There are many native plants growing at the edges of fields, along hedgerows and, perhaps, in your garden that chickens find appetizing. However, always check that they have not been sprayed with pesticides and are not contaminated by exhaust fumes from cars and lorries.

Some of these foragings can be immediately washed and fed to chickens. Chickens usually avoid nettles growing in their runs, or in fields if they are kept as free-rangers, but when boiled nettles become tasty treats.

The range of field, hedgerow and garden-foraged wild plants is extensive – here are a few your chickens will like.

- **Annual Stinging Nettle (*Urtica urens*)**: annual plant, with nettle-like leaves on plants 10–45 cm (4–18 in) high plants. Feed boiled young and tender shoot tips and leaves to chickens.

- **Dandelion (*Taraxacum officinale*)**: perennial plant, widely seen in meadows, pastures, wasteland and gardens. Light green, deeply indented leaves on plants up to 50 cm (20 in) high. Use leaves when fresh and green.

- **Fat Hen (*Chenopodium album*)**: leafy annual, up to 45 cm (18 in) high. Found on wasteland, cultivated soil and along the edges of fields. Pale greyish-green leaves, usually lance-shaped but sometimes variable in outline. Use them when fresh and green.

- **Groundsel (*Senecio vulgaris*)**: an annual mainly growing on disturbed and cultivated soil, where it reaches 10–38 cm (4–15 in) high. Its deep green, succulent, oak-leaf shaped leaves tightly clasp stems. Use when fresh and green.

- **Perennial Stinging Nettle (*Urtica dioica*)**: perennial nature, often with wide-spreading roots. Deep green, nettle-like leaves on plants 30–90 cm (1–3 ft) high. Feed boiled young and tender shoot tips and leaves to chickens.

Groundsel (Senecio vulgaris) *has plenty of taste and eye appeal for chickens.*

Wild fruits and nut

Hedgerow and countryside trees, shrubs and brambles abound in fruits and nuts that chickens find tasty. Here are a few of them.

- **Acorns (*Quercus* spp.)**: gather them – usually means picking them off the ground – in autumn. Use them dried and crushed.

- **Beech nuts (*Fagus sylvatica*)**: known as 'masts' and can be collected in autumn. Use them dried and crushed.

- **Horse Chestnuts (*Aesculus hippocastanum*)**: nut known as a conker that is popular with schoolchildren. Use them dried and crushed. Thoroughly cook them to remove the tannin.

- **Sweet Chestnut (*Castanea sativa*)**: triangular, bright red-brown nuts, borne in two or threes and each 18–30 mm (¾–1¼ in) across and enclosed in pale-green husks. These husks are often known as burs and covered with prickles. Use the nuts dried and crushed.

- **Hedgerow fruits**: late summer and autumn are popular months for picking berried fruits such as blackberries. Use them fresh and sparingly.

Types of food

There are several ways to feed chickens and each has its advantages and disadvantages.

● **Mash**: A balanced food mixture that can be 'dry' or 'wet'. Dry mash is often used where poultry are kept intensively. In this state, the mash keeps the birds busy at a time when their activities are restricted and they have nothing else to do. If fed dry, fresh water must be readily available at all time.

Wet mash includes dry mash plus chopped vegetable waste from kitchens, mixed with hot water. Incidentally, adding water to dry mash does not improve its nutritional value, nor its appeal to chickens, but it is usually more digestible than dry mash.

● **Pellets**: Mixtures of food that have been ground and compressed into cylindrical form. They are easy to handle and use. Additionally, their size can be tailored to suit the poultry.

For easy feeding, pellets can be given to chickens through self-feeding devices, making it an ideal way for feeding chickens if you are away from home most of the day.

● **Crumbs**: These come in a granular form and are popular for feeding chicks (see page 150 for details of their use).

● **Grit**: This is essential for a chicken's digestive process (see page 84 for details). Although food can be ground in a gizzard by normal muscular action, it is not so efficient as when grit is present, which increases the number of grinding surfaces. Additionally, it enables better and fuller action of the chicken's digestive juices.

Crushed oyster shell and limestone grit must be available. Baked and finely ground-up egg shells can be added as a supplement, but ensure they are sterile and not recognizable as parts of eggs, as this may encourage the birds to peck at eggs they or other hens have laid.

Mixtures of food, including oats and barley, help to keep chickens healthy and fit.

The amount of food

This varies and depends on the size of the bird – for example:

★ **Large breed**: Needs 100–150 g (4–6 oz) of food each day.
★ **Bantam-sized**: Needs 50–75 g (2–3 oz) of food each day.

Feeding chicks

For details of feeding chicks, from hatching to 16 weeks old, see pages 150–151.

Cleaning out chicken sheds

Thorough and regular cleaning of chicken sheds and runs is essential. If left, manure encourages the presence of diseases and parasites that infest chickens. Chicken manure can be added to a compost heap for either digging into the soil later or for use as a mulch.

Benefits of chicken manure

- Poultry manure is richer in sulphate of ammonia, phosphate and potash than manure from horses, cattle and pigs.

- Poultry manure contains more than twice the amount of nitrogen as that from pigs, and about four times more than that from cattle.

- The phosphate value is more than twice that expected from pig manure, and more than five times that of cattle manure.

- It is also rich in potash. It has more than twice the amount contained in pig manure and slightly more than there is in cattle manure.

Nature of poultry manure

The high content of nitrogen, phosphate and potash in poultry manure derives from the large amount of plant food that rapidly passes through a chicken's digestive system and is subsequently excreted.

Poultry, unlike mammals, do not defecate and urinate separately. These basic bodily functions are combined and the semi-solid urine (white part of their droppings) can be easily seen; if this – while still fresh – comes into contact with plants it invariably causes damage to leaves and stems. For this reason, chicken manure cannot be put directly on the soil and must first be added to a compost heap.

Poultry manure helps to enrich compost heaps and increase the fertility of garden soil.

The presence of chicken manure in a compost heap encourages other materials to decompose rapidly.

Using chicken manure

The amount of manure expected from a chicken each day is often more than 112 g (4 oz). This means that in a single year just five chickens will produce 206 kg (456 lb) of manure. It is essential to be prepared for this amount of manure and to obtain the best possible results from it.

If just scattered in a heap in a corner of your garden, its fertilizer value is soon leached away by the rain. Here is the best way to use it:

- **Do not allow chicken manure** to become wet, as this makes it difficult to spread.

- **Add it to your compost heap** in 2.5–5 cm (1–2 in) thick layers between topsoil and soft garden waste, such as grass cuttings and vegetable waste. Use only topsoil, as this contains more bacteria and other soil organisms than subsoil.

- **Chicken manure is an excellent activator** in the decay of other materials in the heap.

- **After about a year in a compost heap**, the decayed vegetable compost can be dug into the soil, in autumn or winter. Alternatively, it can be used as a mulch around plants but make quite sure it is not in contact with stems as it may damage them.

- **When used during autumn or winter digging** – and if the soil requires a dusting of lime to make it less acidic – do not apply lime at the same time. Instead, dig the soil and mix in the decayed compost. Then, two to three months later, dust the soil with lime.

Dealing with old poultry manure

Always wear overalls, rubber gloves, a face mask and goggles when cleaning out old chicken sheds and runs. This is because aged and dry poultry manure sometimes contains incubating spores of a human respiratory disease.

If your skin becomes contaminated, you must wash it immediately in hot, soapy water to which a disinfectant has been added. If, by chance, your eyes become contaminated, visit your doctor immediately.

Leafy vegetables especially benefit from the high nitrogen content of poultry manure.

HANDLING AND INSPECTING CHICKENS

Handling a chicken

Chickens are usually friendly and quiet, enjoying the company of people and often becoming so tame they can be treated as pets. Some breeds may even respond to their names. They enjoy being picked up, handled and talked to in a soft and gentle manner. However, take care when picking up young chicks as they may suddenly jump down and harm themselves.

Children and young chicks are a natural partnership, but youngsters should handle chicks carefully, preferably at ground level.

Catching and picking up a chicken

Chickens become skittish and worried if chased. It is far better to train them to come to you; a few food pellets soon do the trick and it is amazing how quickly they associate your arrival with a another meal.

Take care, when catching a chicken, not to grab at their wings. Firmly but gently catch the whole body with both hands.

• **Panicking a chicken** by frantically pursuing it reduces its egg-laying ability, and is certainly bad for its general health. Indeed, loud noises diminish its desire to lay eggs. Additionally, raucous and unexpected noise – from dogs as well as fireworks – may result in misshapen and soft-shelled eggs.

• **Try to usher the hen** into a corner before attempting to pick her up. Do not grab for a wing or tail feathers. Once the chicken is directly in front of you, bend down slowly and pick her up; at the same time, hug the wings close to the body. This secures them and makes it difficult for the chicken to flap her wings and suddenly jump down, which may result in falling on a hard surface. Heavy breeds are especially at risk if they fall from waist height.

• **Once a hen is safely in your arms**, hold her securely and talk to her quietly. If the chicken defecates on you, it is a sign

Chickens delight in being picked up, stroked and talked to in a gentle, soothing tone.

that it is frightened – take care not to drop it in horror, as this is just another aspect of keeping chickens.

• **To carry a hen**, put a hand under its rear end – to hold it secure – and tuck its head under your arm. The bird's head should be slightly lower than its rear end.

Some chicken enthusiasts prefer to have the chicken's head facing towards the front, but this does not enable the bird's vision to be slightly obstructed. A chicken with its sight partially restricted is less likely to become frightened.

• **Never pick up a chicken by its feet or neck** as this causes mental stress and physical damage.

• **Never be tempted to hypnotize** a chicken by putting it on its back; this may lead to the bird suffering heart failure, especially if it is a heavy breed.

Inspecting a chicken

At some time during its life a chicken needs to be carefully inspected, perhaps for external parasites, or because of laying problems or broodiness.

- **Holding a chicken securely** during its examination is essential. Put a hand under the bird, with your middle two fingers positioned between her legs. This firmly secures the bird and gives her confidence at being handled.

- **A chicken is best inspected** towards the end of the day and slightly before dusk, but when there is still good light. This especially applies when checking the bird's crop after she has eaten. If the crop is full, you are providing enough food, but if it is empty give her extra rations during the following weeks. Usually, you will be able to feel food (especially grains and pellets) in her crop, which is in her breast area.

- **Generally, the bird's body needs** to be firm to the touch, without any heaviness or flabbiness. If the bird's body feels slightly 'spare', this may be the result of her recently laying eggs.

- **Check for external parasites** (see pages 166–168 for details of them) by carefully drawing apart the bird's feathers and checking their undersides. Also inspect areas of bare skin. If fleas, lice, mites and ticks are allowed to reach epidemic proportions, they are not only difficult to eradicate but the bird will suffer and egg-laying may decrease.

- **Finding out if a hen is laying eggs** is easily undertaken. If you can place four fingers horizontally between the pelvic bones and the end of the breastbone, she is sufficiently mature to lay eggs. The pelvic bones can be felt on each side of the vent area.

The pelvic bones need to be sufficiently pliable to enable an egg to be expelled and if there is space for only one finger between them the bird is not laying. However, when two or three fingers can be positioned between them, the hen is well able to expel an egg.

Another way to establish if a hen is laying is to check her vent area. In a non-laying hen, the area is yellowish, small and round. In pullets approaching their laying time, the area enlarges slightly and becomes white. Later, the area becomes larger and conical. For a hen in full lay the area is moist and enlarged.

- **Check on broodiness** – a broody hen fluffs up her feathers, squawks and is reluctant to leave a clutch of eggs.

Make sure the chicken feels secure when being held.

With room to exercise, a diet supplemented by foraged food and the opportunity to breed, chickens are less likely to become ill.

PRODUCING EGGS

About eggs

A chicken's egg is a miracle of nature. Primarily it is part of its reproduction cycle but, increasingly, each day eggs are eaten by millions of people throughout the world. There are estimates of billions of chickens in the world at any one time and therefore they are unrivalled food providers.

Which came first?

One of the world's oldest riddles – *which came first, the chicken or the egg?* – has been resolved. A team from two British universities using a super computer has concluded that the 'chicken came first'.

What is an egg?

An average-sized egg weighs about 50 g (2 oz); the shell forms 11 per cent of this weight, the white part 58 per cent and the yolk 31 per cent.

The shell of an egg is a unique piece of engineering, and if today's engineers could construct something as strong and light it would revolutionize the way many things are made.

An egg is formed of many parts and all within a shell that both contains and protects an embryonic chick; also, it withstands stresses incurred when expelled by a hen.

Within an egg there are two easily recognizable parts – the white and yellow elements. But there are many others:

● **Shell**: Usually about 0.3 mm thick and contains an embryo chick in a protective casing. Made almost entirely of calcium carbonate crystals, it is grainy and slightly bumpy in texture. Under a microscope it can be seen to be covered with up to 17,000 pores. It is semi-permeable and enables moisture and air to pass through.

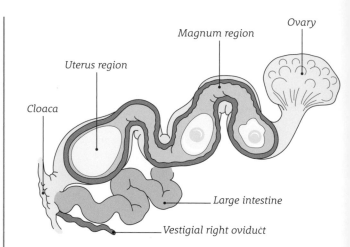

The egg-laying ability of hens is remarkable in its simplicity, enabling a hen to lay both fertilized and unfertilized eggs (depending on whether it has mated with a cock-bird). Eggs develop in the oviduct and are released through the cloaca.

● **Cuticle**: This covers the shell and is a thin layer of organic material that has a distinctive bloom, helping to seal the shell against moisture loss. It also keeps out bacteria and fine dust.

● **Albumen**: This is the so-called 'white' part of an egg which, in four alternate thick and thin layers, surrounds and protects the yolk. It has a nutritional value in the development of chicks and contains about 40 different proteins.

• **Inner and outer membranes**: These are positioned between the shell and the albumen; they are two transparent protein membranes, formed of keratin, that provide defence against bacterial invasion. For this reason, eggs are an incomparable food in regions where perfect hygiene is not possible, especially after natural disasters when water supplies become contaminated with germs.

• **Air space**: Sometimes known as the 'air cell' or 'air pocket', it forms when an egg cools after it is laid. It rests between the inner and outer membranes and is positioned at the largest and widest end of the egg. As an egg ages, this air space slowly enlarges and therefore provides an indication of its age. This air space can be detected when an egg is held against a strong light.

• **Yolk**: This is the central part of an egg and usually yellow, although the precise shade is influenced by the breed of chicken and the food it is given. The yolk contains less water but more protein than the white part of an egg. It has some fat and most of the vitamins and minerals that form an egg. These include iron, vitamin A, vitamin D, phosphorus, calcium, thiamine and riboflavin. The yolk is also a source of lecithin, an emulsifier.

• **Vitelline membrane**: This is a clear membrane that encases the yolk.

• **Chalaza**: The plural is chalazae, and these play an important role within an egg. There are two of them, positioned at opposite ends of the yolk. They anchor the yolk to the inner ends of an egg, thereby keeping it in place. They prevent the yolk becoming damaged if the egg is moved roughly.

• **Blastodisc**: On the surface of the yolk is an area known as the germinal disc: in unfertilized eggs this is called the blastodisc, while in fertilized eggs it becomes the blastoderm and plays an essential role in the progression to the next generation.

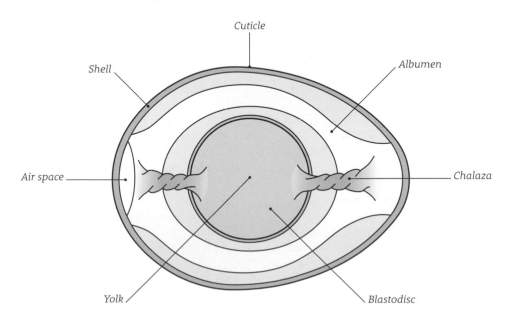

Cuticle

Shell

Albumen

Air space

Chalaza

Yolk

Blastodisc

Eggs are superbly shaped and engineered for rigidity when being expelled by a hen.

Why do we eat eggs?

The most obvious answer is because eggs are an unrivalled and nourishing food. They are versatile in their uses, from egg sandwiches to soufflés and omelettes, with the bonus of being storable for several weeks (see page 102 for storage details). Also, the food within the shell is clean – and in many countries and places this is a bonus.

Here are a few considerations that will put eggs firmly in your diet.

● **Medical research suggests** that eating eggs may prevent age-related macular degeneration of the eyes.

● **Eggs are useful for people who are dieting**. This is because to lose weight the number of calories consumed must be reduced, at the same time maintaining a balanced and nutritional diet. Eggs enable this to be done successfully as they are high in nutritional value but relatively low in calories.

● **Depending on its size, each egg** provides 60–80 calories.

● **Eggs are the best available 'fast food'** and certainly more healthy than highly processed fast foods bought ready for eating. Also, they are less expensive.

● **When eaten as quickly prepared meals** there is a wide range of ways to include them in your own and your family's diet; some of these are featured on page 110 in 'Fast meals for busy families'.

● **Eggs are also ideal for creating family lunches** and suppers and a range of these is detailed on pages 111–129 in 'Egg recipes for all the family'.

Buying, collecting and selling eggs

Buying eggs from a local grocery store or supermarket is a routine most people have experienced. There is also the case of buying eggs from a neighbour who has a surplus, as well as selling your own. Here are a few considerations.

Buying eggs from a grocery store or supermarket

● **Always buy eggs from a supplier** who sells a large number of them each day – and check they are have a long period before becoming 'out of date'. This date is usually three weeks from when the eggs were laid. However, in some countries it is from being packaged, not laid.

● **Always open the lid of the box** and check that the eggs are not cracked and leaking. At first glance this may not be apparent, but by gently trying to move each egg you will soon know if one has leaked – leaking eggs stick to the box.

● **The weight of an egg** is an indication of its freshness. When an egg is fresh, the air cell within it is small. As an egg ages it loses moisture, which is replaced by air. This means that the

Hens are claimed to puff up and peck if they see you removing their eggs.

Eggs need to be cleaned and a colander is ideal for collecting and washing them.

egg becomes lighter in weight. Indeed, there is a quick test that reveals if eggs are fresh. Just put them in a bowl of water – aged ones float, while fresh ones sink to the bottom.

Collecting eggs from your own hens

Commercially, eggs are gathered regularly throughout the day, but for home hen keepers this is usually only practical in the morning and evening. Collecting eggs soon after they have been laid has the advantage in giving hens little time to peck at them.

After collecting eggs from nesting boxes, some chicken enthusiasts prefer to leave the eggs for a couple of days before eating them. This allows the white part of an egg to settle.

To ensure that eggs are eaten in the same sequence they are laid, either place the eggs in dated boxes or lightly write the date on the shell as soon as collected.

Selling your own eggs

Regulations for selling spare eggs to friends or through local shops are decided by national and regional health and welfare agencies. In Europe, the sale of eggs must be compliant with the EU Welfare of Laying Hens Directive. In Britain, the Department for Environment, Food and Rural Affairs (DEFRA) has this responsibility.

These rules and regulations are complex and, although it is not impossible to sell your spare eggs, be aware that the application and paperwork may be complex and visits can be expected from ministry officials. Graded eggs sold at retail level within the EU are legally required to be marked with a code that identifies the egg-laying establishment, country of origin and method of production (free-range, organic, barn or cage). Ungraded eggs sold direct to the final customer are exempt from these markings.

The best advice for home egg producers who wish to sell a few eggs is first to contact DEFRA.

Before selling eggs, check that it is legal (see above).

Winter and summer egg laying

Do not expect your hens to lay as many eggs in winter as they do in summer. Increasing daylight in late spring and early summer stimulates egg production. Conversely, in autumn you will see a decrease in the number of eggs your chickens produce. However, spring-hatched chicks, which usually start to lay eggs in autumn, often continue into winter.

Hybrid breeds are less likely than pure-bred types to have a decrease in their egg-laying abilities, but even these will show a decline after their first year of laying.

Commercial egg producers know that hens normally need a minimum of 16 hours of daylight each day to encourage them to lay eggs. For this reason, additional light is often provided. Indeed, a few 25-watt bulbs are usually enough to encourage egg laying. However, if this is undertaken extra warmth in winter is also needed, and this is usually difficult for home poultry enthusiasts to provide.

Home hen keepers can take advantage of better light in late winter and early spring by orientating a hen house so that it captures the maximum amount of light and warmth (see page 73 for details).

Abnormal eggs

Occasionally, hens produce eggs that have unusual characteristics. These include:

• **Double yolks**: This is quite common and not a problem in deciding whether or not to eat an egg. They are just as good to eat as single-yolk eggs. Double-yolk eggs often occur in hens in their second season of laying. Additionally, some breeds are prone to this peculiarity.

• **Eggs with blood spots**: These are quite harmless to eat and thought to be a hereditary problem. Commercial egg producers 'candle' (see page 157) their eggs and remove any that are showing signs of this problem.

• **Wind eggs**: These are eggs that do not contain a yolk. Young pullets when just starting to lay eggs are prone to this problem. Conversely, a hen which is coming to the end of its egg-laying life may lay wind eggs.

• **Soft-shelled eggs**: The shells are soft and inclined to break. This arises when there is a deficiency of calcium in a hen's diet; add oystershell grit to their food. Loud and unexpected noises can also trigger this problem.

• **Fertile eggs**: These are fine to eat, but some people object to them. They come from a hen that has been fertilized by a cock-bird. Commercial egg producers are usually required by egg regulations to produce eggs with a 'yolk that is free of foreign bodies'. However, if you have a cock-bird in with your hens you must expect to have fertile eggs.

Storing eggs at home

Whether eggs have been bought from a local grocer or supermarket, or are from your own hens, it is essential to store them properly if they are to remain in good condition.

Egg boxes are ideal for storing eggs as each egg is individually and securely supported.

- **When you get your eggs home** – or after gathering them from your own hens – they need to be stored, pointed-end downwards, in a cool place. Shop-bought eggs are invariably packaged and boxed with the pointed end downward (this enables them to fit snugly in the box).

- **Eggs are best left in their boxes**, rather than being placed in an egg-rack. This is because egg-racks are usually positioned close to an ice-box, which is too cold for storing eggs. Also, eggs remain more secure when in their boxes than if placed in a door. Another reason is that when a refrigerator door is opened it enables warm air to circulate around door-stored eggs.

- **An even temperature of 4°C (40°F)** is best and at that temperature you can expect your eggs to remain in good condition for three to four weeks.

- **Storing eggs in a cool room or larder** is often recommended, but invariably this decreases an egg's storage life. Eggs deteriorate faster in warm temperatures than cold ones – 24 hours of storage in a cool room will deteriorate an egg at the same rate as four to five days in a refrigerator. Therefore, eggs stored in a cool room or larder should be used within ten days.

- **Eggs have thin, absorbent shells** so do not put them next to fish, onions or other strong-smelling foods.

- **Where possible, take eggs out** of a refrigerator 20–30 minutes before using them – otherwise, they are more likely to crack on being boiled, or be difficult to whip. Alternatively, after taking eggs out of a refrigerator place them in a bowl of slightly warm water for several minutes.

Freezing eggs

Eggs cannot be frozen whole because the shells break. That means you must either separate the whites and yolks and freeze them individually, or stir the two together and freeze as a whole.

- **When mixed and stored together**, first beat them until well blended, then place in an airtight plastic container. Label and date the container, together with the number of eggs it contains.

- **Egg whites on their own can be beaten and stored** in the same way as suggested above. However, yolks require extra care and need to be beaten in a bowl; if, later, they are to be used in a savoury recipe, add a pinch of salt. Alternatively, use a pinch of sugar if the eventual intention is to use them in a sweet dish.

- **Cooked egg dishes** are difficult to freeze successfully. This is because hard-boiled eggs become rubbery and most quiche or custard type meals tend to 'separate' when thawed.

Did you know

★ Weight for weight, a small egg is of equal nutritional value to a large one.

★ The colour of an egg's shell only reflects the breed of chicken and does not affect its flavour or food value.

★ Immediately after being boiled, hard-boiled eggs should be shelled under cold, running water to prevent discoloration around the yolk, which is harmless but unsightly. Shelling eggs under cold water has another advantage – it enables you to handle a hot egg without your fingers being burned!

★ The texture of an egg is affected if it is cooked at an excessively high temperature. Eggs should be cooked as slowly as possible.

★ A hard-boiled egg turns leathery when overcooked.

Your egg-producing team

The range of hens to consider as members of your dedicated team of egg-layers is described on pages 28–61. Some are also suitable as birds to kill and eat, and these are clearly indicated. Sometimes, when a hen's egg-laying ability radically diminishes, it is worth considering killing and eating her (details of ending the life of a hen are given on pages 136–137).

Some breeds lay more eggs than others – see opposite page for recommended egg-laying breeds.

Looking after your hens and ensuring they lay the maximum number of eggs is often a puzzle for new poultry keepers. Here are a few often-raised questions:

● **How many eggs can I expect my hens to lay?** This is mainly influenced by the breed and some are well known for their egg-laying abilities (see box, opposite).

If you wish to have chickens that are both good at laying eggs and ideal for the table these are known as general-purpose breeds (see box, opposite).

● **At what age can I expect a pullet to start laying eggs?** A pullet is a female chicken less than one year old that has not started to lay eggs. They usually start laying eggs when aged about 21 weeks. Sometimes egg laying happens earlier than this time, and occasionally later.

Occasionally, a pullet has difficulty in laying her first egg and this may result from egg binding (see page 178 for details of this problem).

● **Does a pullet give any sign that she is about to lay eggs?** A pullet nearing her egg-laying time displays a bright red and full comb. You will also notice a general broadening (sometimes known as 'filling out') of her body, especially on either side of her 'vent'. This is the orifice in her backside from where eggs are laid.

A pullet when reaching her time to lay eggs becomes restless, moving in and out of the hen house and clucking in a disturbed manner.

● **How can I encourage a pullet to lay eggs?** If she is in with a group of other hens that are established layers, there will be several nesting boxes present for them to lay eggs in. If the pullet is on her own, put a nesting box in a quiet and dark corner of the house, and place clean, dry straw in it to create a comfortable nesting area.

Sometimes, a pullet will lay an egg on the floor. This must be immediately removed and the area thoroughly cleaned. If left, it will induce her to lay further eggs on the floor. Placing replica china eggs in the nesting box will encourage her to lay eggs in the correct place.

Egg-laying breeds

There are several breeds, including:

★ Barred Rock (see page 33)

★ Black Australorp (see page 34)

★ Cuckoo Maran (see page 40)

★ Faverolles (see page 41)

★ Leghorn (see page 43)

★ Plymouth Rock (see page 46)

★ Rhode Island Red (see page 49)

★ Wyandotte (see page 36)

General-purpose breeds

There are several suitable breeds, including:

★ Barred Rock (see page 33)

★ Black Australorp (see page 34)

★ Buff Orpington (see page 37)

★ Buff Sussex (see page 37)

★ Maran (see page 40)

★ Plymouth Rock (see page 46)

★ Rhode Island Red (see page 49)

★ Wyandotte (see page 36)

Regularly collect eggs from nesting boxes to prevent them being pecked and damaged.

Putting a male bird in with your hens

Having a male bird (known as a cock-bird, cock or rooster) in with your hens is not essential for the production of eggs for eating, and there is no evidence that having a male bird in with your hens will result in them producing any more eggs. However, if you wish to breed a few chickens to raise yourself, a male bird is essential (see pages 24 and 147).

Many breeds are sufficiently docile to be kept as family pets – as well as for laying eggs.

Here are a few advantages and disadvantages to consider:

Advantages

• If you wish to have both eggs for eating and a few for hatching, a male bird is essential.

• The number of 'fertile' eggs increases the longer he is in with your hens.

• Male birds help to keep hens under control, preventing squabbles breaking out between them.

• His presence helps to keep hens contented and placid.

Disadvantages

• The greatest disadvantage, especially in a suburban garden, is the noise he creates. Hens on their own are not a problem, but when a male bird is introduced the noise may disturb the entire neighbourhood – especially early in the morning!

• It is essential to buy a male bird from a reputable source, so that his pedigree is known.

• A cock-bird introduced into a group of hens which have never seen one before can be disturbing for them, and it may even put them off laying eggs.

Cock-birds are usually either killed and eaten, or selected for their ability to mate with hens.

● De-spurring is usually essential to prevent him damaging hens during mating (see page 147 for details).

● Eggs may be fertile (see page 102).

● Cost of buying a male bird.

● Cost of feeding.

● He needs to be mature – ten or more months old.

● The fertility of a cock-bird usually decreases after 5–6 years, so be prepared to replace him after that period.

● He should have been vaccinated against Fowl Pest (see page 172). This is also known as Newcastle Disease and is a notifiable problem (see page 165).

● If you wish him to fertilize hens and to produce chicks, make sure that he is not closely related to them.

EGG RECIPES

Cooking with eggs

Eggs are so popular and versatile that they are not always given the respect they deserve. They are excellent ingredients for use in 'fast' meals, as well as in the creation of family lunches and suppers. Here are a few of the best ways you can use them.

Fast meals for busy families

There are many ways to use eggs in fast meals, ranging from boiled or fried for breakfast to egg sandwiches and omelettes. Here is a parade of fast egg meals – they are fun as well as nutritious and fortifying.

● **Egg sandwich**: Basically, this is a soft fried egg between two pieces of buttered bread. It is quick and easy to make, and in military circles it is known as an 'egg banjo'.

To make another form of egg sandwich, finely chop hard-boiled eggs, add seasonings and, perhaps, mayonnaise, and spread the mixture between two pieces of buttered bread.

● **Boiled egg**: This needs little description but care is needed to ensure success. Getting the desired degree of 'softness' or 'hardness' needs experimentation and most cooks have their own ideas about the best method to produce a soft-boiled egg. One way is to put an egg into cold water in a saucepan, heat until boiling, then time it for four minutes. This produces an egg with the yolk soft and slightly runny. Always boil an egg gently, so that the shell does not crack.

Serve the egg with toast or buttered bread. Dipping toast or bread soldiers into the yolk has great appeal for children!

● **Scrambled eggs**: This is another well-known way to cook eggs. Allow one or two eggs for each person. Crack open each egg, put into a bowl and beat lightly (if desired, add seasonings such as salt and pepper). To cook four to six eggs, put 25 g (1 oz) of butter in a saucepan and when melted add the egg mixture. Cook gently and stir with a wooden spoon until stiff, then place on top of pieces of hot buttered toast.

● **Poached eggs**: Quick and easy and can be cooked in an egg poacher or pan of boiling water. Poached eggs are usually served on hot buttered toast, with a dusting of pepper to add further flavour.

● **Omelettes**: These have many devotees and their versatility – with fillings ranging from cheese, tomatoes and ham to jam and fish – is legendary, and all within an egg mixture coaxed into firmness in a flat-based, non-stick pan over a low heat.

Basically, for each person whisk two or three eggs in a basin and add to a pan in which 25 g (1 oz) of butter has been gently melted. On a low heat, repeatedly use a broad-ended, non-stick spatula or the end of a 'turner' to keep the edges of the mixture from burning. Keep 'fluffing' the sides inwards. When the mixture starts to become firm, add a desired ingredient on one side of it. Then, turn the other half of the mixture over and on top. After a few minutes turn the omelette completely over to ensure each side is lightly and evenly brown. Serve on its own or perhaps with a light salad.

To create a main meal using an omelette as a basic part, consider adding a salad of lettuce, cucumber and tomatoes. Potato salad is another way to add flavour. These combinations result in some of the best slimming meals, providing all the nutrition you need for a healthy life.

Egg recipes for all the family

Eggs can be combined with many other ingredients to create delicious meals that both adults and children can enjoy. Here are a few interesting ways to use them in main meals – the mouthwatering recipes in this section include Curried Eggs, Egg and Bacon Bubble, Onion Quiche, and Tomato Egg Mash.

Kedgeree

Fish and rice and all things nice, including eggs, make up this tempting kedgeree. A breakfast that's good for supper too.

- Serves: 4
- Preparation: 20 minutes
- Cooking: 20 minutes

Kedgeree

You'll need

4 eggs
175 g (6 oz) long-grain rice
175 g (6 oz) smoked haddock
50 g (2 oz) butter
Salt and pepper
1 beaten egg (optional)
Chopped parsley

How to prepare and cook

1 Hard-boil the eggs and remove the shells under cold running water. Slice one egg and chop the rest.

2 Cook the rice in boiling water for about 15 minutes, until tender. Drain well and separate the grains.

3 Cook the haddock in simmering water for about 10 minutes, then drain. Melt the butter in a saucepan, add the rice and fish, and stir with a fork until hot. Mix in the chopped eggs and seasoning. Optional: Stir in the beaten egg to make the mixture creamy. Serve hot and garnished with sliced egg and chopped parsley.

To freeze: This recipe is not suitable for freezing.

Egg and Sardine Slice

This is the type of meal that looks as if lots of preparation has gone into it, yet it is really quite simple to make.

- Serves: 4–6
- Preparation: 25 minutes
- Cooking: 15 minutes

You'll need

6 eggs
1 tin boneless sardines
Juice of half a lemon
15 ml (1 tbsp) chopped parsley
Salt and pepper
175 g (6 oz) flaky or puff pastry
1 beaten egg

How to prepare and cook

1 Preheat the oven to Gas Mark 7 (220°C/425°F).

2 Hard-boil the eggs and chop finely. Break the sardines into small pieces. Mix together with the lemon juice, chopped parsley, salt and pepper.

3 Roll out the pastry to an oblong 35 × 30 cm (14 × 12 in). Place the egg and sardine mixture lengthwise on half of the pastry, to within 2.5 cm (1 in) of the edges. Brush the edges with beaten egg, fold over the other half of the pastry and seal the edges. Brush with more beaten egg.

4 Make incisions across the top of the pastry, about 2.5 cm (1 in) apart, and bake for 15–20 minutes.

To freeze: This recipe is not suitable for freezing.

Egg and Sardine Slice

Curried Eggs

Eggs served on a bed of rice and cooked in a rich, hot, curry sauce. They make a warming light lunch or a tasty supper dish.

- Serves: 4
- Preparation and cooking: 40 minutes

You'll need

2 onions, skinned and sliced

3–4 tbsp olive oil

45 ml (3 tbsp) curry powder

15 ml (1 tbsp) plain flour

1 cooking apple, cored and chopped

600 ml (1 pint) stock

15 ml (1 tbsp) garlic chilli sauce

1 tsp chopped root ginger

15 ml (1 tbsp) honey

15 ml (1 tbsp) fine-cut marmalade

Juice of half a lemon

Salt and pepper

60 ml (4 tbsp) double cream

225 g (8 oz) basmati rice

8 eggs

How to prepare and cook

1 Fry the onions in the oil until soft. Stir in the curry powder and flour and cook for 2–3 minutes.

2 Add the chopped apple to the pan. Cook for a further 2–3 minutes. Mix in the stock and bring to the boil, stirring all the time. Reduce the heat and add the garlic chilli sauce, root ginger, honey, marmalade and lemon juice. Season with salt and pepper. Simmer for 25–30 minutes, stirring occasionally. Strain into a clean saucepan, add the cream and leave over a gentle heat.

3 Meanwhile, cook the rice in boiling water for about 10–12 minutes, until tender. Drain well, rinse and fork through to separate the grains. Place on a warmed serving dish.

Curried Eggs

4 Hard-boil the eggs and remove the shells under cold running water, cut in half lengthwise and arrange on the rice. Pour over the sauce. Garnish with fresh parsley or coriander and Greek yoghurt.

To freeze: This recipe is not suitable for freezing.

Egg and Chicken Mousse

This dish presents stuffed olives, sliced eggs and chicken, set in a tasty aspic jelly. A superb starter for summer days.

- Serves: 4
- Preparation: 30–40 minutes

You'll need

4 eggs
225 g (½ lb) cooked chicken meat
400 ml (¾ pint) aspic jelly (or gravy from cooked chicken with gelatine added)
15 ml (1 tbsp) sherry
60g (2 oz) stuffed olives
150 ml (¼ pint) double cream (optional)

How to prepare and cook

1 Hard-boil the eggs, then shell under cold running water. Leave them in cold water. Cut the chicken into small pieces.

2 Combine the aspic and the sherry. Wet a 1.1 litre (2 pint) mould. Pour a small amount of aspic into the mould and tilt to coat the bottom and sides. Place in a refrigerator.

3 Slice the olives and arrange them in the mould. Pour another layer of aspic over the olives. Allow to cool.

4 Slice the eggs. Dip each slice in the aspic and place in the mould. Pour a little more aspic over the eggs and return to the refrigerator to set.

5 Add the chicken and top up the mould with aspic. Option: whip the cream and fold in the remaining aspic and chicken. Return to the fridge until set.

6 Turn out and garnish with spring onion or other salad.

To freeze: This recipe is not suitable for freezing.

Egg and Chicken Mousse

Tomato Egg Mash

An interesting meal from storecupboard ingredients – and it is easy to prepare. An attractive meal for all family members.

- Serves: 4
- Preparation: 15–20 minutes
- Cooking: 20 minutes

You'll need

700 g (11/2 lb) potatoes (or a large packet of instant potato)
Salt and pepper
Milk
5 eggs
454 g (16 oz) can tomatoes
30 ml (2 tbsp) sage-and-onion stuffing
75 g (3 oz) cheese (grated)

How to prepare and cook

1 Boil and mash the potatoes. Preheat the oven to Gas Mark 5 (190°C/375°F).

2 Beat the seasonings into the potato, then add a little milk and 1 egg to make a creamy mixture. Place the mixture in a warmed ovenproof dish, making a hollow in the centre.

3 Empty the tomatoes into the middle of the dish, sprinkle with seasonings and the sage-and-onion stuffing mixture. Break the remaining eggs on the tomatoes, cover with cheese and bake for 20–25 minutes, until the eggs are just set.

4 Serve hot and with a green vegetable.

To freeze: This recipe is not suitable for freezing.

Tomato Egg Mash

Onion Quiche

A mouthwatering savoury meal that is equally good hot or cold. Try it with a crisp green salad and French bread.

* Serves: 4
* Preparation: 30 minutes
* Cooking: 35 minutes

You'll need

175 g (6 oz) shortcrust pastry
450 g (1 lb) onions (skinned and sliced)
50 g (2 oz) butter
3 eggs
100 g (4 oz) sliced ham, chopped
300 ml (½ pint) single cream or milk
Salt and black pepper
Large pinch of ground nutmeg

How to prepare and cook

1 Preheat the oven to Gas Mark 6 (200°C/400°F).

2 Roll out the pastry and use it to line a 20 cm (8 in) flan ring; stand it on a baking sheet.

3 Lightly fry the onions in butter until semi-transparent and soft. Cool slightly and place in the flan case with the ham.

4 Beat the eggs and cream or milk together and season with salt, pepper and nutmeg. Pour it over the onions.

5 Bake in the oven for 10 minutes, then reduce the temperature to Gas Mark 4 (180°C/350°F) and cook until set, about 25 minutes. Garnish with fried bacon and spring onions.

To freeze: This recipe is not suitable for freezing.

Onion Quiche

Greek Egg and Lemon Soup

This traditional Greek soup is quick to prepare and has a delicious, subtle flavour.

- Serves: 4–6
- Preparation: 5 minutes
- Cooking: 15 minutes

You'll need

50 g (2 oz) long-grain rice
1.4 litres (2¹/₂ pints) chicken stock
2 eggs
Juice of 2 lemons

How to prepare and cook

1 Wash the rice and place it in a large saucepan. Pour the stock over it.

2 Bring to the boil and simmer gently until the rice is just cooked, about 10–15 minutes. Allow it to cool slightly.

3 Lightly beat the eggs with the lemon juice. Add a few spoonfuls of the stock, stirring well.

4 Gradually add the remaining stock and pour over the rice. Heat very gently, taking care not to boil as the egg may separate.

5 Serve immediately with French bread or Melba toast.

To freeze: This recipe is not suitable for freezing.

Greek Egg and Lemon Soup

Tripledecker Favourite

Smashing for a weekend brunch or when the kids come home starving! It is a variation on the all-time favourite – eggs, bacon and potatoes.

- Serves: 6
- Preparation: 20 minutes
- Cooking: 15 minutes

You'll need

700 g (1¹/₂ lb) potatoes (peeled and cooked)
Salt and black pepper
60 ml (4 tbsp) plain flour
Oil for frying
6 rashers of bacon
6 eggs

How to prepare and cook

1 Mash the potatoes and season well. Add the flour and mix well to form a stiff dough. Add more flour if necessary. Roll out on a floured surface to 5 mm (¹/₄ in) thick. Cut out 12 rounds, using a 7.5 cm (3 in) wide cutter.

2 Heat the oil in a frying pan and fry the potato cakes until golden-brown on both sides. Place them on absorbent paper and keep warm.

3 Fry the bacon and eggs. Place one rasher and one egg on half of the potato cakes. Use the other cakes to form lids for them.

To freeze: This recipe is not suitable for freezing.

Tripledecker Favourite

Anglesey Eggs

This is a meal with a special Welsh flavour – where the taste of leeks is combined with an egg and cheese sauce.

- Serves: 4
- Preparation: 30 minutes
- Cooking: 20 minutes

You'll need

700 g (1½ lb) mashing potatoes (peeled)
6 medium-sized leeks
Knob of butter
Salt and pepper
8 eggs

Sauce

25 g (1 oz) butter
25 g (1 oz) plain flour
300 ml (½ pint) milk
75 g (3 oz) grated Cheddar cheese
Pinch of dry mustard

How to prepare and cook

1 Preheat the oven to Gas Mark 6 (200°C/400°F).

2 Boil the potatoes, strain and mash. Clean the leeks well and cut into rings. Boil for 10 minutes. Drain and add to the potatoes, together with butter and seasonings. Beat well to give a fluffy mixture. Place it in a warm, ovenproof dish and keep warm.

3 Hard-boil the eggs, remove the shells under cold running water, then cut in half and place on top of the potato mixture. Keep this warm while making the sauce.

4 Melt the butter in a small pan, remove from the heat and add the flour. Gradually add the milk, return the pan to the heat and bring to the boil, stirring all the time. Cook for one minute. Add 50 g (2 oz) of the cheese, seasonings and mustard.

5 Coat the eggs with sauce. Sprinkle the remaining cheese over the top and place in the oven for 20 minutes.

To freeze: This recipe is not suitable for freezing.

Anglesey Eggs

Eggs Divan

Try this classic American dish on your family – they will love it. You can use two eggs for each portion, or one egg and leftover chicken.

- Serves: 4
- Preparation: 35 minutes
- Cooking: 20 minutes

Eggs Divan

You'll need

8 eggs, or 4 eggs and some leftover cooked chicken
450 g (1 lb) cooked broccoli, asparagus or green beans

Stuffing

Salt and cayenne pepper
Mayonnaise

Sauce

40 g (1¹/₂ oz) butter
25 g (1 oz) plain flour
300 ml (¹/₂ pint) stock or milk
100 g (4 oz) cheese (grated)
Salt, pepper and made-up mustard

How to prepare and cook

1 Preheat the oven to Gas Mark 6 (200°C/400°F).

2 Hard-boil the eggs, shell them under cold running water and cut a slice lengthwise from one side of each of them. Scoop out the yolk and mash well with salt, cayenne pepper and enough mayonnaise to moisten. Spoon the mixture back into each egg and replace the earlier cut slices as lids.

3 To make the sauce, melt the butter in a small pan, remove from the heat and add the flour. Gradually add the stock or milk, return the pan to the heat and bring to the boil, stirring all the time. Cook for one minute. Add three-quarters of the cheese, seasonings and mustard to taste.

4 Arrange the broccoli, asparagus or beans in the buttered, ovenproof dish and cover with some of the sauce. Put the eggs and chicken (if used) on top. Pour the remaining sauce over the top and sprinkle with the remaining cheese.

5 Bake in an oven until golden-brown (usually takes about 20 minutes).

To freeze: This recipe is not suitable for freezing.

Croque with Eggs

The traditional French Croque is a hot, fried sandwich with a variety of fillings. Serve as a snack, light lunch or supper.

- Serves: 6
- Preparation: 10 minutes
- Cooking: 15 minutes

You'll need

6 slices of bread, about 5 mm (¼ in) thick
Butter
Dijon mustard
6 slices of Cheddar cheese
6 eggs
Salt and black pepper

How to prepare and cook

1 Preheat the oven to Gas Mark 5 (190°C/375°F).

2 Trim the crusts from the bread. Spread the bread with butter and mustard. Place in a buttered, shallow, ovenproof dish.

3 Place a slice of cheese on each slice of bread. Put the dish in the oven until the cheese begins to melt.

4 Carefully break an egg on top of each slice. Season well and return to the oven until the egg has just set (usually about 10 minutes). Serve at once.

To freeze: This recipe is not suitable for freezing.

Croque with Eggs

Oeufs en cocotte

This very simple and delicious dish takes its name from the small individual dishes the eggs are cooked in. It makes a splendid starter.

- Serves: 6
- Preparation: 5 minutes
- Cooking: 10 minutes

You'll need

50 g (2 oz) butter
6 large eggs
90 ml (6 tbsp) cream
Salt and black pepper

How to prepare and cook

1 Preheat the oven to Gas Mark 4 (180°C/350°F). Melt the butter and use to thoroughly grease the inside of 6 ramekin dishes. Place the dishes on a baking tray.

2 Carefully break an egg into each dish. Warm the cream slightly and pour 15 ml (1 tbsp) over each egg. Season. Bake in an oven for 5–10 minutes, until just set – the yolks should still be runny. Serve at once with fingers of hot toast.

Variations

- Sprinkle a small quantity of fresh chopped herbs on the top of each egg.
- Place finely chopped ham or chicken in the base of each of the dishes.
- Sprinkle the top of each egg with grated cheese.
- Lightly fry chopped onions and mushrooms and place in the base of each dish.
- Place flaked tuna fish in the base of each dish before adding the eggs.

To freeze: This recipe is not suitable for freezing.

Oeufs en Cocotte

Egg and Bacon Pasties

The basic ingredients are the traditional British breakfast – eggs and bacon – but this recipe is slightly different and can be served hot or cold.

- Serves: 4
- Preparation: 20 minutes
- Cooking: 20 minutes

You'll need

7 eggs
100 g (4 oz) bacon rashers (rinds removed)
20 g (³/₄ oz) fresh breadcrumbs
Salt and pepper
175 g (6 oz) shortcrust pastry

How to prepare and cook

1 Preheat the oven to Gas Mark 6 (200°C/400°F).

2 Hard-boil 6 of the eggs, shell them under cold running water, and chop them up. Chop the bacon. Beat the remaining egg and mix with the chopped eggs, bacon, breadcrumbs and seasonings.

3 Divide the pastry into four. Roll out into larger circles. Dampen the edges of the pastry and pile the egg mixture to one side of each of them. Fold the pastry over, seal and flute the edges together.

4 Bake near the top of the oven for 15–20 minutes. Serve hot or cold.

To freeze: Allow the pasties to cool. Pack carefully in rigid containers. Label and seal and freeze them rapidly. They can be frozen for up to two months. To serve, allow them to thaw at room temperature for two to four hours. If desired, reheat them in an oven at Gas Mark 6 (200°C/400°F) for 15 minutes.

Egg and Bacon Pasties

Egg and Bacon Bubble

This is a substantial family supper, a complete meal in itself. It is attractive and contains family favourites.

- Serves: 4
- Preparation: 20 minutes
- Cooking: 40–45 minutes

You'll need

2 eggs
30 ml (2 tbsp) oil
450 g (1 lb) potatoes (cooked and cubed)
1 onion (skinned and chopped)
4 rashers bacon (rind removed and chopped)

Custard

2 eggs
300 ml (½ pint) milk
Salt and pepper

How to prepare and cook

1 Preheat the oven to Gas Mark 5 (190°C/375°F).

2 Hard-boil the eggs, remove the shells under cold running water, then leave in the water. When cool, chop them up.

3 Heat the oil and fry the potatoes, onion and bacon together for 5 minutes. Drain and place in a shallow, 1.1 litre (2 pint) ovenproof dish, together with the chopped egg.

4 Beat the eggs, milk and seasonings together and pour over the mixture.

5 Bake in the oven until the custard is set, about 35 minutes. Serve immediately, while still hot.

To freeze: This recipe is not suitable for freezing.

Egg and Bacon Bubble

Bread-and-Butter Cheese Pudding

Economical, nourishing and very convenient – it uses storecupboard ingredients. A recipe with great family appeal.

- Serves: 4–6
- Preparation: 15 minutes
- Cooking: 35–40 minutes

You'll need

8 slices of bread (can be slightly stale)
50 g (2 oz) butter
100 g (4 oz) cheese (grated)
4 eggs
568 ml (1 pint) milk
Salt and pepper

How to prepare and cook

1 Preheat the oven to Gas Mark 5 (190°C/375°F).

2 Remove the crusts and spread the bread with butter. In a 1.1 litre (2 pint), well-buttered ovenproof dish, place layers of bread and grated cheese.

3 Lightly beat the eggs, milk and seasonings together and pour over the bread.

4 Bake for about 35 minutes, until set and golden brown. Serve at once.

To freeze: This recipe is not suitable for freezing.

Bread-and-Butter Cheese Pudding

Soufflé Omelette

Fluffy Soufflé Omelettes can be given many flavourings and fillings. No matter how often you serve them, there is always variety.

- Serves: 1
- Preparation: 10 minutes
- Cooking: 10 minutes

You'll need

2 standard-sized eggs
10 ml (2 tsp) cold water
30 ml (2 tbsp) caster sugar
12 g (½ oz) butter
Icing sugar (to dust)

How to prepare and cook

1 Put a pan over a low heat to become hot. Meanwhile, separate the eggs and whisk the yolks in the cold water and sugar until pale and creamy. Whisk the whites until just stiff, then fold into the yolk mixture. Prepare the grill.

2 Melt the butter in the pan, then pour in the omelette mixture. Cook this mixture, without moving it, until its base is set and golden. Then, place under a hot grill for about half a minute.

3 Fold the omelette in half and slide onto a warm plate; dust the top with icing sugar. Serve immediately.

Flavourings

Add one of these to the whisked egg yolks:
- Grated rind of a lemon or orange.
- 50 g (2 oz) of plain chocolate melted in 10 ml (2 tbsp) of water (allow to cool slightly before adding to the eggs) and remember to omit the water in the basic recipe.

Fillings

Add along the line of the fold of a cooked omelette 30 ml (2 tbsp) of any of the following (first slightly warmed):
- Ginger
- Orange jam
- Marmalade
- Fruit purée

To freeze: This recipe is not suitable for freezing.

Caption

Eggs Baked in Tomatoes

This simple and colourful snack is quickly prepared and cooked. Large tomatoes are the most suitable for this recipe.

- Serves: 4
- Preparation: 10 minutes
- Cooking: 10 minutes

You'll need

4 large tomatoes
Salt and black pepper
1 small onion (skinned and chopped)
25 g (1 oz) butter
4 small eggs
4 croutons of fried bread

How to prepare and cook

1 Preheat the oven to Gas Mark 5 (190°C/375°F).

2 Wash the tomatoes and use a sharp knife to cut off their tops. Use a small spoon to carefully scoop out the pulp from the insides of the tomatoes. Sprinkle the inside of each tomato with salt and pepper. Place them on a baking tray.

3 Lightly fry the onion in butter until soft and place a spoonful in each tomato.

4 Carefully break an egg into each tomato. Season to taste.

5 Bake until just set, about 10 minutes. Serve immediately on fried bread or croutons.

To freeze: This recipe is not suitable for freezing.

Eggs Baked in Tomatoes

Egg and Cauliflower Mimosa

An interesting dish – eggs and cauliflower with a tangy redcurrant or cranberry sauce.

- Serves: 4
- Preparation: 25 minutes
- Cooking: 20 minutes

You'll need

One medium-sized cauliflower
5 eggs
75–100 g (3–4 oz) bacon rashers (rind removed and chopped)
25 g (1 oz) butter
12 g (½ oz) plain flour
300 ml (½ pint) stock
Salt and pepper
30 ml (2 tbsp) redcurrant or cranberry jelly

How to prepare and cook

1 Break the cauliflower into small sprigs, add the green leaves and cook in boiling water until just tender. Strain and measure the cauliflower water for the sauce. Spread the cauliflower in a buttered dish and keep hot.

Egg and Cauliflower Mimosa

2 Meanwhile, hard-boil one of the eggs, shell under cold running water and chop the white part, adding it to the cauliflower in the dish.

3 Fry the bacon in butter until crisp. Mix in the flour and, when bubbling, stir in the stock, seasonings and jelly. Stir the mixture until thickened and cooked – keep the sauce hot.

4 Poach the remaining eggs, drain on absorbent paper and arrange on top of the cauliflower. Pour hot sauce over the top and rub the hard-boiled yolk through a sieve to garnish the surface. Serve hot with potatoes or toast.

To freeze: This recipe is not suitable for freezing.

La Piperade

This is a simple yet nourishing dish from the Basque region of Europe. Serve it as a light lunch or supper dish.

- Serves: 4
- Preparation: 15 minutes
- Cooking: 15 minutes

You'll need

45 ml (3 tbsp) oil
1 large onion (skinned and sliced)
1 clove garlic (skinned and crushed)
2 red peppers (seeded and chopped)
450 g (1 lb) tomatoes (skinned and chopped)
Pinch of dried basil
Salt and black pepper
6 eggs (beaten)
15 ml (1 tbsp) parsley (chopped)

How to prepare and cook

1 Heat the oil in a large, heavy frying pan. Add the onions and garlic and fry gently until soft.

2 Add the peppers, tomatoes and basil and cook until tender and almost pulped, about 10 minutes. Season to taste.

La Piperade

3 Add the beaten eggs and cook, stirring all the time, until they are just beginning to thicken. Remove them from the heat.

4 Sprinkle parsley on top and serve while still hot.

To freeze: This recipe is not suitable for freezing.

Egg and Anchovy Croquettes

This recipe is perfect for a light snack. The croquettes can also be served as an accompaniment to a light meal.

- Serves: 4
- Preparation: 30 minutes
- Cooking: 5–7 minutes

You'll need
4 eggs
50 g (2 oz) tin anchovies
25 g (1 oz) butter
25 g (1 oz) plain flour
150 ml (¼ pint) milk
Black pepper
100 g (4 oz) Cheddar cheese (grated)
Beaten egg and white breadcrumbs for coating
Oil for deep frying

How to prepare and cook
1 Hard-boil the eggs and remove the shells under cold running water. Allow them to become cold, then chop finely.

2 Drain the anchovies and finely chop.

3 Melt the butter in a pan, remove from the heat and stir in the flour. Gradually add the milk. Return to the heat and bring to the boil, stirring all the time. Cook for one minute. Add the pepper and cheese, mixing well.

4 Add the eggs and anchovies and allow to become quite cold. When cold, shape into cork-shaped pieces. Coat with egg and breadcrumbs. Heat the oil until a 25 mm (1 in) piece of bread will become brown in a minute. Fry the croquettes until golden-brown, about 5 minutes. Drain well on absorbent paper and serve immediately.

To freeze: This recipe is not suitable for freezing.

Egg and Anchovy Croquettes

CHICKENS FOR THE TABLE

Chickens for meat

There are several sources of chickens for eating and these include buying ready-to-eat commercially raised and fattened hens, eating your own hens when they reach the end of their economical egg-laying lives, and buying pullets (young hens) which you can fatten up yourself. Each of these has different welfare considerations.

Chickens for eating

There are several options and each of them has advantages and disadvantages:

● **Buying from a supermarket**: Ensure the chicken has been raised in a non-stressful manner and not in battery conditions, where it will have lived in an area less than the size of an A4 piece of paper (see page 66 for details of the exact space allotted to each chicken); its quality of life will have been pitifully low. On the plus side – if there is one – such hens are the least expensive of bought, ready-to-cook chickens. But this is meaningless to a chicken that has lived its life in something resembling a penal colony.

● **Buying from farm shops and local butchers**: This enables customers to directly ask proprietors and butchers to detail the sources of their chickens and how they were raised.

● **Capons**: This is a term for a male chicken that has been neutered by surgery or chemicals to reduce its growing and fattening time and to improve the eating qualities of its flesh (see page 133 for details).

Alternatives to surgery and chemicals to neuter chickens are possible (these are also described on page 133).

● **Eating your own chickens**: The egg-laying abilities of your hens eventually decreases, raising the question whether it is economical to continue feeding them. It is a difficult decision to make about killing chickens you have got to know for several years. Additionally, it introduces questions about the technique of killing (see pages 136–137), plucking (see pages 138–139) and drawing and trussing (see pages 142–143).

Even if you kill a chicken – and do not eat it – there are strict regulations about disposing of the body (see page 137).

● **Raising your own chickens for eating**: This has become increasingly popular, with young pullets (often at the age of five weeks) being bought, raised and eaten 10–12 weeks later (see pages 144 onwards for details).

Raising chickens for their meat is increasingly popular.

'Boiler' hens and capons

Invariably, the flesh on hens killed after their egg-laying abilities decrease is tough and claimed to be only suitable for *coq au vin*. Nevertheless, such birds are edible but need careful preparation and a long cooking time; after killing, plucking and evisceration (clearing out a bird's internal organs) the body is simmered in hot water for several hours before being placed in an oven for cooking. Such preparation helps to tenderize the flesh and make it more flavoursome.

Eating and preparing a 'boiler' (an old hen) in this way has been part of country living for many centuries, creating a nourishing meal in stews and casseroles and, especially, with the addition of root vegetables. As a family meal it should not be discounted in favour of the 'trendy' and more modern approach to eating chickens.

Incidentally, do not confuse a 'boiler' with a 'broiler', which is a young bird (usually less than eight weeks old) solely raised to be killed for its meat. Usually, this is undertaken commercially and, in some instances, a bird weighing 2 kg (4¹/₂ lb) can be produced in five weeks.

What is a capon?

The history of capon chickens dates back to the Romans, who are credited with their introduction. About 160 BCE, as a way to conserve grain rations, the fattening of hens for eating was prohibited. As a solution, they castrated young male birds which, at the end of their usual growing time, resulted in the doubling of their body size. Increasingly, capons became popular, with so-called farmyard chickens being regarded as only fit for peasants to eat. Prejudice against eating farmyard chickens was further increased by monks proclaiming the virtues of capons.

• **Caponizing a young male bird**: Traditionally, caponizing involved making an incision in the bird's side – between the last two ribs and close to the tail – and removing the testicles.

This surgery involved a certain amount of cruelty so using chemicals to neuter male chickens was tried. The female hormone oestrogen was inserted in pellet form under skin in the bird's neck. It resulted in the shrinking of the bird's testicles and reversal of its sexual characteristics.

• **Banning chemicals and surgery**: Both of these neutering techniques have been banned in many countries. The use of chemicals puts eaters at risk of absorbing female hormones, while surgery causes unnecessary distress to a chicken.

• **Alternatives to surgery and the use of chemicals**: There are two better and more humane methods of encouraging male chickens to grow rapidly. These are:

Buy young, hybrid birds that have been specially bred for the production of meat for the table.

Separate male birds from the females when 12–14 weeks old and place them in a small pen on their own. They must not be allowed to roam freely. The fattening and fast growth required from them is produced by feeding them more or less 'on demand'. This means feeding them three times a day with a mixture of equal parts oats, barley and boiled potatoes, with the addition of a little skimmed milk. Give the birds as much of this mixture as they can eat; ensure they are also given plenty of clean water. Birds treated in this way can often be killed and eaten about three weeks later.

Breeds of chicken for meat and egg production

Choosing a breed of chicken that will fulfil its purpose is essential. Some breeds are known for their egg-producing qualities, while others can produce both eggs and meat (see page 105). There are also others that are better famed for their meat-producing ability.

Ross Cobb chickens

These are mainly commercial broilers (young birds, usually less than eight weeks old, specially raised to be killed for their meat). There are several forms but generally they have broad chests and large feet to support their bodyweight.

The white-feathered variety is popular, with a voracious appetite and rapid growth rate. Commercially, they reach the saleable size of 2 kg (4 lb) in five weeks, but home poultry keepers take 10–12 weeks to achieve this size. However, they have been known to quadruple their size in six weeks.

Some birds develop so rapidly that they begin to have thoughts of adulthood and produce early notes of a 'cock-a-doodle-doo'. If this stage is reached, they need to be killed and eaten.

Young birds of this breed are available from most poultry suppliers.

Popular meat (table) breeds

- Croad Langshan (see page 40).

- Red Dorking (see page 47).

- Rhode Island Red (see page 49).

Popular egg and meat (dual-purpose) breeds

- Barred Rock (see page 33).

- Black Australorp (see page 34).

- Buff Orpington (see page 37).

- Buff Sussex (see page 37).

- Maran (see page 40).

- Plymouth Rock (see page 46).

- Rhode Island Red (see page 49).

- Wyandotte (see page 36).

Please note: Within many of these breeds there are colour variations, but the nature of them remains the same. For example, within the Wyandotte breed there are four colours.

Squabbles sometimes arise between chickens – see page 17 for details. For the benefit of the bird's health and for the purpose of keeping healthy and productive chickens, it is sensible to allow them as much space as possible.

Killing chickens

Few other tasks associated with keeping poultry cause such anxiety as having to kill a chicken. This may be necessary because birds earlier bought to be fattened are ready for eating, or old hens have reached the end of their egg-producing lives and need culling. Also, a chicken sometimes has to be killed because it is infirm and needs to be put out of its misery.

The legalities of killing a chicken

The process of killing a chicken must be humane, causing the least stress. It also has to be easy for the 'terminator' as if complicated the bird may suffer unnecessarily. There is also the complication that home poultry keepers may not have experience in dispatching a chicken. If this is so, ask for advice from a local hen-keeping society.

A practical and traditional way to kill a chicken is described on this and the following page.

Questions are often raised about killing, eating and selling your own chickens: these can be complicated and directives about them vary from one country to another. However, the following general advice will help:

● **Can I eat one of my own chickens that I have killed?** The answer is yes, but the chicken must have been slaughtered humanely and lawfully.

● **What are the slaughtering options?** There are two ways available to you. Kill your own chickens as described here, or arrange for an approved slaughterhouse to kill them on your behalf.

It is unlawful to have your chickens killed anywhere else.

● **Can I sell meat from my own chickens?** No – this would be illegal. Only if the chicken was killed in an approved slaughterhouse would it be possible to sell the meat. Additionally, unless the chicken is killed in a slaughterhouse it would be illegal even to use the meat to feed paying guests in a bed-and-breakfast enterprise owned by you.

Methods of killing chickens

Commercially, killing a chicken is usually performed by a stunning device. However, home poultry keepers have to rely on the traditional method which causes death through neck dislocation. This is also known as neck pulling. When done correctly, the bird immediately and irrecoverably loses consciousness.

How to kill a chicken

This is not difficult, but if you are in any doubt about your ability to undertake the task, *do not attempt it*. Ask an experience poultry keeper to do the dispatching for you or give guidance about it. Also, local chicken clubs will be able to help with advice, or enquire about a 'poultry course' in your area.

Killing a chicken is not for a person with a sensitive stomach.

● **Do not feed the chicken for 24 hours** before it is to be killed. However, ensure it has plenty to drink as this will help to keep it calm.

- **As well as killing the chicken**, care must be taken not to stress the remaining ones. Select the chicken to be killed and take it to another enclosure or, preferably, a quiet shed so that the bird is on its own with you.

- **Putting the bird in a large, straw-filled box** and covering it with a cloth helps to keep it calm. Alternatively, provide the bird with a perch about 30 cm (1 ft) high. This helps to keep the bird in one position.

- **How the chicken is held** is influenced by its size and the dispatcher's height. If you are short in stature, stand on a firm, slightly raised surface so that the bird's body hangs freely to one side and without touching the ground.

Do not stand on a small-surfaced, rickety box that rocks and may cause you to fall off if you unexpectedly step backwards. Where possible, stand on the ground.

- **Hold the chicken with your weaker hand** (the left one if you are right-handed, and vice versa), with its legs upside down and firmly clasped. Hold the legs with your thumb pointing away from your body; this enables a firm grip to be made.

- **Ensure the chicken's body is hanging down** and across the front of your body.

- **With your stronger hand, hold the neck** so that at least three strong fingers and your thumb are around it; then turn the neck to a horizontal position. Always use your strong hand and arm to undertake the dislocation.

- **With a sudden pull downwards**, dislocate the chicken's neck. This should leave a gap between the neck and the last vertebra. It is better to pull downwards too firmly than not to break the neck. If the head comes away from the neck, this can be messy but is much better than not completing the task and leaving the chicken partly alive.

- **The bird often flutters its wings** and moves its legs for a few moments after being dispatched; this is quite natural, even though it is dead. During this period, continue to hold the legs and neck firmly.

- The dead bird can be suspended by its feet from a strong hook in a cool and shaded place.

Method of killing a chicken.

> ## Disposing of a dead chicken
>
> If you intend to eat a chicken after killing it, disposing of the body is not a problem. However, if you decide not to eat it, the body must be disposed of legally. This also applies to chickens killed in road accidents, through old age or as a result of illness.
>
> As soon as you have a dead chicken and wish to dispose of it, contact your local Animal Health Office, an organization that gives advice about dealing with dead animals. In most areas, the organization that can assist you is the National Fallen Stock Company.

Plucking chickens

Removing all of a chicken's feathers, known as plucking, is essential, either immediately after death or when the body is cold (see below for details). It enables the chicken's exterior to be thoroughly cleaned and dried, as well as allowing the body to be prepared for other pre-cooking tasks, such as hanging, drawing (evisceration) and trussing.

How soon after death should I pluck a chicken?

There are two schools of thought about the best time to pluck a chicken:

- **A chicken's body remains warm and floppy** for about 45 minutes after death and before it stiffens as a result of rigor mortis. Most poultry enthusiasts prefer to pluck a chicken during this period, when feathers come out easily.

- **Plucking can also be undertaken** when the body is totally cold. If plucked when only slightly warm, there is a likelihood of the skin being torn and damaged.

Scalding a chicken

This is a technique recommended by some poultry keepers as an aid in the removal of feathers, especially if the chicken's body is cold. It involves dipping the chicken in hot water, about 62–65°C (145–150°F). The water should not be excessively hot, as you will need to put your hands in it.

Hot water softens and melts fat around the bases of feathers, enabling them to be easily pulled out. The technique of scalding is:

- **Hold the chicken by its feet** and dip the body and legs into the water for 3–5 seconds. It is essential to jiggle and swish the body to ensure that the hot water penetrates the bases of the feathers.

- **Completely remove the chicken's body** and, after a few seconds, replace it in the water, again for 3–5 seconds.

- **Remove the chicken and tug** at one of its large feathers. If it comes out easily, scalding is complete and successful. If, however, feathers are still held firmly, repeat the dipping and jiggling process. If necessary, undertake this process until feathers can be easily pulled out.

Plucking a chicken

Experienced chicken enthusiasts are forever telling novice hen keepers how they can pluck a chicken in five or less minutes. This may be so, but in reality and for new chicken keepers the time taken will be far greater. Here is practical guidance on plucking a chicken.

- **Select a well-illuminated shed** (so that you can clearly see what you are doing) or a secluded place outside that is not exposed to strong, gusty wind. When blown about, small, fluffy feathers are difficult to collect.

- **Wear a face mask**, especially if you are working in fully enclosed conditions where dust might be raised from the feathers.

• **Where possible, spread a large, old sheet** or piece of plastic where plucking will take place, to collect the feathers.

• **Suspend the chicken's body by its legs** from a strong hook, so that it is level with your shoulders; this makes it easy to see and reach the feathers. Alternatively, lower the body so that you can sit on a stool or chair yet still be able to reach the feathers. Positioning the chicken's body is a matter of choice, but it must be comfortable for you.

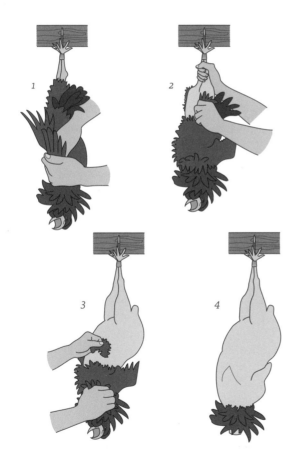

The stages of plucking a chicken (see text, top right).

• **It is essential that feathers are pulled out** from the bird's body in the same direction from which they emerge from the flesh. Do not twist or pull them at a different angle as the flesh will be damaged. Always follow the sequence outlined below (see illustration, bottom left):

1 Start by plucking the primary feathers. These are the long, stiff, flight feathers found at the outer tip of each wing. Hold several feathers together and give them a sharp tug.

2 Next, pluck the leg feathers, taking care not to tear the flesh.

3 From the legs, work down the body, systematically removing all the feathers.

4 Eventually, the bird's body and legs will be entirely free from feathers.

• **Cut off the bird's neck** at the point of dislocation. The body then can be 'drawn' (see pages 140–141 for details).

Dealing with plucked feathers

Feathers have a high nitrogen content and are ideal for adding, in thin layers, to a compost heap. Thin, fluffy ones decay faster than thick, quill-like types, which may take two or more years to decay. This is not a problem because, when you are removing decomposed compost from a heap or bin, feathers that have not fully decayed can be replaced in another heap for a further year.

The decayed garden compost is ideal for digging into the soil in autumn or early winter. Such compost is also good for adding to the sowing or planting positions of runner and French beans.

Hanging and drawing a chicken

Whether the body of a chicken needs hanging is a matter of personal choice – there are many opinions about the necessity of this procedure, but the purpose is to reduce the core temperature to 0°C (32°F).

Hanging a chicken

Here are a few facts to consider:

● **The texture of the flesh** improves with the onset of natural breakdown.

● **The flavour of the meat** is improved (but care must be taken not to hang the chicken for an excessively long period).

● **Fats in the chicken's body** solidify.

● **The soft tissue within the body** firms up, aiding its removal (see 'Drawing a chicken', right).

Hung chicken.

> ### How to hang a chicken
> This is done by suspending the chicken in a cool, clean room or vermin-proof shed, with a good circulation of air and temperature under 3°C (37°F), for 24 hours.
>
> An alternative method is to place the chicken in a refrigerator. Tie a plastic bag over the chicken's head to catch drips of blood from the nostril and beak. If the head and part of the neck have been removed, tie a bag over the end.

Drawing a chicken

It is essential to 'draw' (also known as 'eviscerate' or 'gut') a chicken. This entails the removal of the bird's 'innards'. These include the heart, liver, outer case of the gizzard (part of the stomach, where the bird grinds its food), and the remaining part of the neck. These are known as the 'giblets' and, occasionally, are retained for simmering into a nourishing and traditional broth.

The technique of drawing

It is essential to undertake this task on a clean, recently scrubbed, germ-free surface. Additionally, at the end of this process the surface must be scrubbed and disinfected.

Ensure you have a sharp knife and several clean bowls readily available; you may also wish to wear an apron and thin rubber gloves. Here is a step-by-step guide to drawing a chicken (see illustrations on opposite page):

1 Place the chicken on a firm surface; hold the body securely. Then, using a sharp knife or poultry secateurs, cut the neck to about 2.5 cm (1 in) above the bird's shoulders. Additionally, slit the neck further down, cutting through the muscle so that it breaks away easily.

2 Peel away the crop (the portion of the chicken's food channel, where food is stored until ready for passing into the stomach) and cut it out close to the neck cavity. Insert a finger into the neck cavity to loosen the ligaments close to the breastbone; prise them away and remove.

If you intend to use the gizzard in the preparation of food, clean it by using a sharp knife to cut halfway through it lengthwise, then pull out the yellow bag of grit and throw it away.

3 Turn the bird so that its rear end faces you. Use a sharp knife to form a wedge-shaped cut to sever the fatty oil sac (sometimes known as the 'parson's nose') on the rear end of the bird and positioned close to its tail. If this is left, it may give a peculiar taste to the meat.

4 Make a further cut, between the parson's nose and the vent (this is the orifice through which the chicken urinates, defecates and expels eggs). Carefully cut around this area, taking care not to pierce the rectum, then pull out the intestines.

5 Remove all of the innards, which includes the liver, heart and kidneys.

Lightly wash then thoroughly dry the inside. Sometimes, all that is necessary is to wipe the inside with a damp cloth.

Bacteria present inside the chicken will be killed during cooking, which must be thorough.

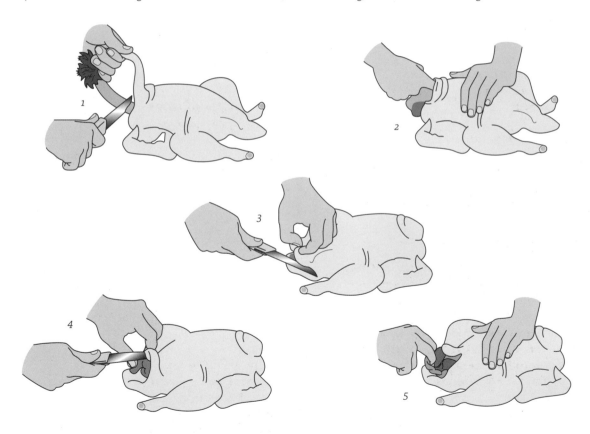

The technique of drawing a chicken (see text, opposite and above).

Trussing and freezing a chicken

Tidying up and trussing a chicken after it has been eviscerated is sometimes considered to be part of that particular task (see pages 140–141), but to many hen-keeping enthusiasts it is a separate step in the preparation of a bird for cooking. If a chicken is left unsecured and not trussed, it makes cooking difficult.

A succulent, home-reared chicken has plenty of eye-appeal – and taste!

Trussing and singeing

When all the innards have been removed, the chicken can be trussed.

- **First, fold back loose pieces of neck skin** over the hole created by the removal of the neck. Place the bird on its back and pass a piece of strong thread across the thighs and above the hocks (the leg joints, between the lower leg and thigh).

- **Pass the ends of the string** up and alongside the body. Tie them firmly.

- **Use another piece of string** to secure the legs in place, although sometimes a single piece is used to hold both the legs and thighs in position.

- **Sometimes, there are small, unsightly hairs** on a chicken's body. Fortunately, they can be quickly removed by singeing. Move a burning piece of tightly folded paper over the hairs. Alternatively, hold the carcass over a lighted gas jet, but take care not to burn and blacken the skin. Using long matches is another way to singe hairs.

Some chicken enthusiasts like to singe the hairs before the bird is fully trussed. However, waiting until after trussing prevents legs and thighs flopping about and becoming damaged in a flame.

Freezing and defrosting

After a chicken has been drawn (see pages 140–141), with its innards removed, it should be cooked and eaten as soon as possible. In the meantime – and for a short period – place it in a refrigerator, but away from uncovered and uncooked food.

If cooking and eating the chicken immediately are not an option, it can be put in a home deep-freezer.

- **Freezing**: Place the chicken inside a moisture-resistant wrapper, with its giblets in a separate bag. The temperature within the deep-freezer must, initially, be set to its lowest point to enable the chicken to lose temperature rapidly. This ensures that the meat retains its flavour.

After the chicken has been in a deep-freezer for two or three days, the temperature can be raised to its normal settings.

- **Defrosting**: A frozen chicken must be completely thawed before cooking, and for the best results this needs to be done slowly; a rapid temperature change on thawing will damage the flesh.

Once a chicken has been thawed, it is as perishable as fresh meat and should be kept in a cold place or refrigerator and cooked as soon as possible. Do not refreeze once thawed.

Indication of the time required to thaw a chicken

Weight		At room temperature 18°C (65°F)	In a refrigerator 4°C (40°F)
kg	lb	hours	hours
0.9	2	8	28
1.3	3	9	32
1.8	4	10	38
2.3	5	12	44
2.7	6	14	50
3.2	7	16	56

- **Defrosting in a microwave**: This is possible, but not as satisfactory as putting a frozen chicken in a refrigerator or placing it at room temperature.

A chicken defrosted in a microwave must be cooked immediately afterwards. This is because part of the chicken may have become warm during the defrosting process and started to cook. This would encourage the presence of bacteria and the chance of food poisoning occurring.

RAISING YOUR OWN CHICKENS

Raising chickens

Few poultry-keeping sights are as idyllic as a cock-bird strutting among his ladies, with chicks following in convoy behind mum. It is nature's way of ensuring succession from one generation to another. If you do not want a cock-bird in your group of hens, young chicks can be bought when a few weeks old.

Why replace existing hens?

Many chickens are kept for their egg-producing qualities and a range of them is detailed on pages 28–61. Invariably, the ability of a hen to lay eggs diminishes after its second year of laying and, eventually, it becomes uneconomical to keep her. There are several ways to replace her, including:

• **Buying young chicks that you can raise** until they become pullets and start to lay eggs (this usually takes 20–22 weeks). See pages 150–151 for details of raising them and the type of food they require.

• **Buying point-of-lay (known as POL) pullets** to replace hens which are coming to the end of their economical life. This is an efficient replacement method as aged hens can be eaten (see pages 136–143 for details of humane killing and preparing them for eating).

If you decide not to eat the hen, after killing it the body must be disposed of in a legal manner (see page 137 for details); bodies cannot just be buried on your own land.

• **Breeding your own chicks** as replacements for aged hens. For this to be successful, healthy hens are essential. The characteristics and egg-laying abilities of a wide range of breeds is detailed on pages 28–61. Additionally, the desired characteristics in a hen and a cock-bird are described on the opposite page.

Incidentally, it is better for home poultry keepers to concentrate on breeding from pure breeds than to have thoughts about producing new hybrids.

The bucolic scene of young chicks clustering around a mother hen appeals to many poultry enthusiasts.

Choosing birds for breeding

You may have an interest in raising chickens from a breed you admire. Alternatively, you may already have a hen you wish to mother future chicks. Before selecting a hen for breeding purposes, carefully inspect her with a clinical eye, not an emotional one. If a particular hen is friendly and becomes a family pet, this is not sufficient for her to be chosen for breeding. Choosing the right cock-bird is equally imporant.

Desired hen characteristics
Remember that it costs just as much to feed a poor-laying hen as one with a good egg-laying record. Here are a few things for you to consider:

● **She should be sexually mature**: A hen suitable for breeding needs to be in her second year of laying eggs. When a pullet, she should have reached sexual maturity at about 22 weeks. Pullets that do not lay eggs until much later are not suitable as they carry such characteristics into their progeny.

● **Good laying qualities**: To ascertain this from your own hens, keep daily records of egg laying. Also, check on the quality and shape of the eggs – avoid mature hens that have a record of laying misshapen eggs.

● **Breed only from healthy hens** (what to look for is detailed on page 163).

Desired cock-bird characteristics
Cock-birds usually remain sexually active for four years, although some are known to be still functioning at six years of age. Therefore, it is essential to buy a cock-bird from a reputable supplier who can assure you of his credentials, such as coming from a mother with a good egg-laying record so that this can be passed to his progeny. Here are a few important considerations:

● **Ensure he is in good health** (general considerations are detailed on page 163).

● **He should not be closely related** to the hens.

● **He should be at least ten months old**, and preferably slightly more.

Looking after your cock-bird
He needs special attention if he is to remain in good condition for several years and be able to service his hens.

● **A high-protein diet is vital** to ensure he remains in good health; adding chopped liver to his rations helps to keep him pepped up.

● **Cock-birds need regular rests** from their sexual activities. For one or two days each week put him in a separate pen.

● **A heavy-breed cock-bird** should not have more than eight hens to look after – preferably only six. However, a light-breed cock-bird can be allowed to have ten birds.

● **Regularly trimming his spurs** (just above his feet) is essential to ensure that he does not harm a hen when mating with her. Keep the ends of his spurs about 12 mm ($^1/_2$ in) long. This is best done by first wrapping him in a towel and then using wire-cutters for the trimming.

Incubation and rearing chicks

After mating with a cock-bird a few times, a hen will start to lay fertile eggs. She will gather them together under her body to hatch naturally. The temperature under her is about 37.7°C (100°F). Hatching takes about 21 days, although it may vary by a day earlier or later. This is not something to worry about, because nature has to be allowed to take its course and the hatching of eggs cannot be rushed.

A hen that has become broody and willing to sit on a cluster of eggs will exhibit the following behaviour:

- **She will sit on eggs** without any encouragement.

- **If she is approached** and perceives the possibility of being disturbed, she will fluff up her feathers and produce a loud, guttural squawk. This is to tell you to move away from her eggs.

- **If she is moved**, she will walk around, still fluffed up, and make her way back to the eggs.

The brooding cycle

Hens are naturally programmed to lay two clutches of eggs each year, which they sit on in order to produce chicks. The first clutch appears in spring, the second in late summer.

One result of creating hybrid breeds has been that this natural inclination of hens to have two clutches each year has been almost eliminated. Nevertheless, it is a characteristic that remains in pure-bred types. Therefore, if you need a surrogate mother hen to sit on an excessive number of eggs your mother hen has produced, it is to these pure breeds that you must look for assistance. Within the descriptions of breeds of chickens (pages 28–61), information on those which easily become broody and therefore will be good as surrogate mothers can be found.

Finding a comfortable place for a broody hen

A hen, after laying eggs and wishing to hatch them, needs special attention. Here are a few ways to give her the comfort she requires:

- **Gentle warmth and cosiness** are essential. Also, the area must be dry, free from draughts and protected against vermin, especially rats.

- **Ensure the area is free from mites and lice** (see pages 168 and 166 for descriptions of them).

- **Dry and warm bedding** are essential, such as straw, hay or wood shavings.

- **A quiet place, away from other hens,** helps her to relax.

- **Fresh water and food** must be readily available to her, and always close by.

A broody hen's daily movements

These are the routines a broody hen naturally adopts and should not worry you:

- **Each day she usually gets up** and takes some exercise. This is an opportunity for you to check on the eggs.

- **If she does not get up**, gently lift her up and check if all the eggs are still present.

- **It is not necessary for you to turn the eggs**, as she will naturally attend to this job. If an egg is left in one position, the embryo chick eventually sticks to the inside of the shell.

- **During the last three days before hatching**, lift off the hen and sprinkle the eggs with slightly warm water to ensure they are not becoming dry. During this process, the eggs must not be allowed to become cool.

The time to hatch

This is an important time and usually occurs on day 21 (or very close to it) of an egg being laid and nestled under a mother hen. A chick will start to 'pip' (peck) its way out of

The effort needed on the part of a chick to break out of its shell is quite considerable.

the shell. Each chick has a special hard part on its beak to make this possible.

Often, there is a temptation to help a chick that is noticeably struggling to break free, but this should be resisted. Weak chicks, unable to naturally peck free from the shell, will not live long or develop into healthy adults.

However, if the shell is thick and resists widening attempts by a chick – which has already pierced the shell – use slightly warm water to soften it. Do not use water excessively as it may trickle down the chick's nostrils and drown it. Always replace the chick, together with its broken shell, back among the other hatching chicks.

Chick imprinting

An imprinting relationship with a chick begins during its first day of life. If the mother hen is present the chick imprints with her, but if you are the only living creature present (and this may happen if eggs are being hatched in an incubator – see pages 152–157) it will imprint on you and consider you as its leader and protector.

Within the third day from hatching, a chick starts to recognize fear. Normally, the mother hen will calm it, but if it has imprinted on you try talking softly – and even gently singing – to it. Essentially, avoid making sudden movements and loud noises.

Established chicks bought by mail order or by direct collection from a supplier will never imprint with you in the same way as home-raised ones. However, if it becomes the sole member of your flock of chickens it will see you as top member in the pecking order (see page 16 for details).

Looking after hatched chicks

As soon as chicks break out of their shells, they need special attention and feeding to ensure they grow healthily, strongly and rapidly. If they are not given the correct amount and quality of food, as well as warmth, their development is retarded.

Feeding chicks

For the first 24 hours in their lives, chicks do not need any kind of food. Thereafter, and until 16 weeks old (when they can be given an adult diet), they need nutritious, growth-promoting food.

● **Initial feeding and until 6–8 weeks old**: Give them a high-protein proprietary chick feed, usually formed of crumbs (known as 'starter crumbs') which contain all the nutrition they require for a balanced diet.

Initially, these starter crumbs are more digestible to chicks if slightly dampened with milk. A few drops of cod liver oil can be added to aid the chicks' development.

Finely chopped hard-boiled eggs mixed with finely grated wholemeal breadcrumbs is another choice. Clearly, this diet takes more preparation time than just giving the chicks proprietary starter crumbs, but it might occasionally be introduced into their diets as a special treat or if the chicks are losing their appetite for the starter crumbs.

During these early weeks in their lives, the chicks need to be fed regularly and often during daylight hours, usually every four hours. The mother hen will show the chicks how to peck at the starter crumbs, but if they are slow to learn fill a shallow dish with food and tap its sides to gain their attention. There are always a few chicks that are slow to discover food – and they need help.

Warmth, protection and the correct food are essential.

A supply of clean, fresh water is another necessity, using a shallow dish. Do not use a deep and large dish as the chicks may not be able to reach the water. Also, in their youthful excitement they may fall into the water and drown.

● **At 6–8 weeks of age**: Their diet can be slowly changed to 'grower's crumbs'. At first, and for a short period, combine the starter crumbs with the grower's crumbs. It is essential not to change suddenly from one food to another. Throughout this period, keep a keen eye on them to ensure they are eating and growing properly. If the appetite of a chick falters when given grower's crumbs, return to the starter crumbs for 7–10 days. Then repeat the introduction of the grower's crumbs.

Mother hens are havens of tranquillity and safety for young chicks.

● **At 16 weeks**: The chicks can be slowly introduced to an adult, egg-laying mixture formed of pellets or mash (see page 87 for details). If the chicks do not readily take to this adult food, return to the previous food for a few days and give them both types. Then, slowly change to the adult food.

Temperature for chicks

Gentle warmth is essential for chicks during their early development and this needs to compare with that provided by the mother hen when she was broody.

Ensure that the area is free from draughts and temperature fluctuations and is proofed against vermin, especially rats and mice.

Continue providing the mother hen with warm and dry bedding, such as hay, straw or wood shavings. The chicks, especially when very young, will snuggle up to her for warmth as well as a feeling of safety.

Chick feathering

Also known as 'feathering up' and 'feathering out', this occurs when a chick is six to eight weeks old (sometimes slightly later in some breeds). They will have lost their initial fizzy coverings and developed early feathers and, at that stage, be known as 'hardened off'.

If the weather is mild, they can go outside in a wire-netting run with protection from strong sunlight, rain, wind and vermin. A run about 1.8 m (6 ft) long and 60 cm (2 ft) wide and similarly high will accommodate 12 chicks. As well as having space for the chicks, room is needed for feeding and drinking bowls. Ensure the ground is free from parasites. This usually means an area where hens have not recently been kept.

If the weather is cold, a similar area indoors needs to be constructed; this is sometimes known as a 'haybox brooder' (see page 159) and enables chicks to continue growing in comfortable conditions.

Artificial incubation

As an alternative to allowing a mother hen to hatch her own eggs (or for a broody hen to act as a surrogate mother), it is possible to buy fertile eggs and to encourage them to hatch in an incubator. Several easy-to-use types of incubator are available for home poultry keepers.

Broody hen or artificial incubator?

If a mother hen is not able to sit and hatch her own eggs, then the next best way is to use a broody hen. This is the traditional alternative method for encouraging eggs to hatch. If, however, a broody hen is not available, the fall-back position is to use an artificial incubator (see below and opposite for the various types).

Advantages and disadvantages of using an incubator

There are several advantages and disadvantages in using an incubator, but the thrill of seeing chicks emerging from their shells is immeasurable and something that will enthral your entire family.

Advantages
● **Once bought, an incubator allows you** to raise pullets to replace hens which have reached the end of their economical egg-laying lives. Its cost can therefore be recouped over many years. Occasionally, a second-hand incubator can be bought, but you must make absolutely certain that its electrics are safe and suited to your electricity supply.

● **An incubator enables eggs to be hatched** even when a broody hen is not available.

Disadvantages
● **The cost of buying an incubator** may appear high if you intend to use it only a few times. Therefore, it might be cheaper to buy pullets that are about to begin laying eggs. Alternatively, you might be able to share the incubator with hen-keeping friends.

● **The incubator must be installed** in a dry, quiet, vermin-proof shed which has an assured and uninterrupted electricity supply.

● **Eggs must be marked and regularly turned** during their period in an incubator (see page 156). If this is neglected, the embryo chick will eventually stick to the inside of its shell.

● **Space is needed in a dry, clean and airy shed** to keep the incubator when not in use.

● **You will need to closely monitor the temperature** in the incubator for about three weeks (see page 154 for the required temperature).

Types of incubator

These vary in size: many are suited for use by large poultry-keeping establishments, others for more modest demands where only a few eggs need to be hatched each year.

Don't buy a large incubator just because it looks professional. It will be expensive and uneconomical for you when a more modestly sized one will be just as efficient.

There are two main types – 'still-air' and 'forced-air'.

● **Still-air incubators**: This design does not have a fan to circulate air around the eggs to ensure an even and uniform temperature is maintained (see page 154 for details of the desired temperature). Instead, the air stratifies, making it difficult to control or to monitor the actual temperature in the incubator. This design factor can lead to eggs becoming overheated or cold. The best way to assess the temperature in a still-air incubator – and therefore to know whether to lower or raise the temperature – is by using two thermometers and taking an average reading. The temperature-reading part of each thermometer should be about 5 cm (2 in) above the tops of the eggs, and not close to the source of heat.

● **Forced-air incubators**: This design has a built-in fan that circulates the air at the correct temperature. The temperature should be set at 37.2–37.5°C (99–99.5°F). This allows the temperature to rise slightly to the desired 37.7°C (100°F).

To make egg turning easier and more certain, mark each egg (see page 156).

Installing an incubator

For an assured and reliable performance, always buy the best-quality incubator you can afford. Installation must be made by a qualified electrician in a dry, clean and airy shed. Preferably, it must be near your house, so that you can easily check on it throughout the year. A light above the shed's entrance will make the routine of checking easier.

Install the incubator on a level and firm surface, where it cannot be knocked and jarred. Also, a position away from direct and strong sunlight prevents overheating and damage occurring to the eggs.

Health and safety are essential considerations, and therefore you should not have cables strewn over the floor, where you can trip over them. If an incubator is damaged through a cable being inadvertently pulled out, you might lose several weeks' work.

If there is a power failure, keep the eggs warm by placing several blankets over the incubator. Candles put in jars and positioned close to an incubator covered with a blanket will help to maintain an adequate temperature for several hours.

If a power cut lasts more than a few hours, later 'candle' the eggs to check they are still fertile (see page 157).

Preparing eggs for incubation

Ensure that healthy eggs are chosen for incubation. Here are some guidelines to follow:

• **Essentially, eggs must be those laid** by hens which have been with a cock-bird for several weeks. The eggs must also be fertile.

• **Do not select eggs from pullets** that have only just started to lay eggs.

• **Avoid cracked eggs** – hairline cracks may have allowed infection to enter.

• **Eggs should weigh** about 50g (2 oz) – or slightly more.

• **Do not use extra-large eggs** as they could contain two yolks and be unlikely to produce chicks.

• **Fresh eggs have a better chance** of producing healthy chicks than old ones. Therefore, they should not be more than seven days old (see 'Storing eggs before incubation', below).

Storing eggs before incubation

If you are continuously incubating batches of eggs, it may be necessary to store some of them while waiting for space in an incubator to become available.

★ **It is possible to store fertile eggs** for up to 14 days before putting them in an incubator, but preferably this should be for no more than seven days.

★ **Store eggs, pointed-end downwards**, in egg trays in a temperature of 12.7–15.5°C (55–60°F) and with a humidity reading of 70–75 per cent. Temperatures in excess of 23.8°C (75°F) and with humidity less than 40 per cent will damage them.

Using an incubator

The aim within an incubator is to replicate conditions underneath a hen when hatching eggs. Apart from turning the eggs regularly (see page 156), the main requirements are correct temperature, humidity and ventilation.

• **Temperature**: The desired temperature is 37.7°C (100°F) which, ideally, needs to be maintained level with the eggs, not above or below them. Embryos in eggs become damaged – often fatally – if the temperature rises above 39.5°C (103°F) or falls below 35.5°C (96°F) for several hours.

When the incubation period nears its end (about 21 days), regularly check for emerging chicks.

A mother hen naturally maintains the desired temperature, but when using electrical equipment this can be difficult and the temperature needs to be checked several times each day.

● **Humidity**: Preferably, the relative humidity should be about 60 per cent for most of the hatching period, but rising to 65–70 per cent during the last three days: remember that the incubation period is about 21 days.

The relative humidity within an incubator can be influenced by changing the size of the water pan. Should it be necessary, putting a small sponge in the water pan creates an increased area for water evaporation, thereby raising the relative humidity within the incubator.

A wet-bulb thermometer can be used to assess the relative humidity. Remember than any increase in the relative humidity affects the temperature and therefore regularly check both of these readings to maintain the correct temperature and relative humidity.

● **Ventilation**: The eggs need a freely available supply of air, so check the ventilation holes and slots in the incubator are not blocked – but avoid having cold air passing over the eggs.

Putting eggs in an incubator

This is an exciting time in the lives of poultry keepers, an adventure with the prospect of new life being created after about three weeks.

1 Make sure the incubator is clean and that the electrics are working safely. Place it on a level and firm surface.

2 Switch on the incubator two to three hours before it is needed to ensure that the correct temperature is reached by the time you need it.

3 Make sure the water tray (at the base of the incubator) is full of clean water. This is essential to ensure the correct humidity is maintained (see 'Using an incubator', page 154). It will be necessary to regularly top up the water tray.

4 Before putting eggs in an incubator, use a marker pen to put an X on one side and an O on the other side of each egg. This will readily enable you to identify eggs you have turned (see 'Turning the eggs', right).

5 As well as marking eggs with X and O, write the expected hatching date on each egg (allow 21 days from placing in the incubator). Marking eggs with the date is essential if you have several batches of them in the same incubator.

6 Carefully place each egg (on its side) in the incubator. Do not cram a small incubator with masses of eggs; it is better to have fewer and to spread them out.

7 Re-check that the temperature and humidity are correct. After seven or eight days the eggs will need to be 'candled' (see below).

Candling eggs

This a way to check on the inside of an egg and see if it is fertile. It is best undertaken seven or eight days after putting an egg in an incubator.

Turning the eggs

Turn the eggs completely over three or more times each day. The markings made on the shells will enable you to see which eggs have been turned.

★ **Some poultry aficionados recommend** turning each egg an odd number of turns each day; this ensures that an egg does not spend consecutive nights on the same side.

★ **Stop turning eggs** on the 19th day.

★ **Soon after the 19th day** you will be able to hear chicks using their beaks to peck at the insides of their shells. This pecking continues until the chicks break out.

★ **Some incubators are fitted** with a window through which you can see what is happening to the eggs.

● **Candling equipment**: Home-made candling devices are easy to construct and formed of a box-like structure: one example is a wooden box, 15 cm (6 in) square and 20–23 cm (8–9 in) long. In the inside of one end, drill a narrow hole (for a cable to pass through) and secure a bayonet light fitting to the inside, so that a 60-watt light bulb can be inserted in it.

At the opposite end to the light bulb, cut a 3 cm (1¼ in) wide hole. It is through this hole that an egg will be examined. Do not cut the hole on the top surface of the candling device, because heat in the box rises and may damage an egg when it is being inspected.

Proprietary candling equipment is widely available and easy to install.

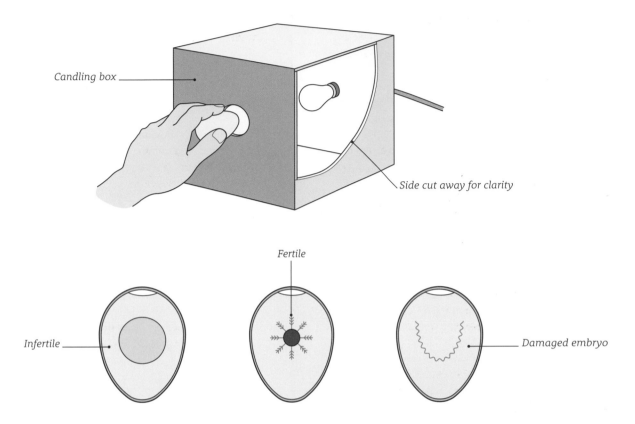

Candling box

Side cut away for clarity

Fertile

Infertile

Damaged embryo

Candling is a quick and easy way to check if an egg is healthy and fertile.

- **How to candle an egg**: Each egg has to be held firmly but carefully and candled individually, in the following way:

- **Turn on the candling box** and turn out all lights in a dark room.

- **Take the eggs, one by one**, from the incubator and position against the light emitted from the hole at one end of the candling box.

- **Look directly down on the egg**, or at a slight angle if this is easier for you.

- **If the egg contains a small, dark red spot** with veins radiating outwards in all directions, then it is fertile and can be returned to its position in the incubator. Candling needs to be done quickly to prevent the egg becoming cold.

- **If the egg has a clear inside**, it is infertile and will not produce a chick. Remove the egg; boil and chop it up for feeding to your hens.

- **Occasionally, you will see** a jagged, irregular line inside the egg; this usually results from the shell being damaged. You should discard such eggs.

Looking after newly hatched chicks

Newly hatched chicks need careful attention to ensure that they develop healthily. However, it is essential to remove weak, blind and deformed chicks as soon as possible (see below). For the healthy ones, you will need to provide either a broody hen or an artificial brooder to keep them warm and safe.

Removing weak and unhealthy chicks

Killing a young chick is not a job for the squeamish, but it has to be done. One way is to break its neck with a stout piece of wood. Another is to sharply tap its head.

Whatever the method used, it is essential to keep young children away from the area when this is being done. The sight of dead chicks, or a parent dispatching them, could be deeply disturbing for youngsters for many years afterwards.

Immediately after hatching

Most home poultry enthusiasts like to leave the young chicks alone for 24 hours after they hatch to enable them to become dry and fluffy. Chicks do not need to be given food during this short period as they will be relying on nourishment they earlier had available to them when in an egg.

There are two choices of what to do with them; sometimes, a broody hen will accept an 'instant family' of young chicks (see opposite page), or they need to be put in an artificial brooder (see below).

Sometimes, newly hatched chicks in an incubator will imprint with the first person they see and who looks after them (see page 149).

Using an artificial brooder

If a broody hen is not available, the chicks have to be put in an artificial brooder. This is a warm, dry, vermin-proof area that restricts the chicks and provides a safe, secure home for them.

A large box, perhaps 90 cm (3 ft) square and with a similar height, is ideal. It needs a flap-like door on one side to enable access and for the chicks to be checked and handled. The base of the flap should not be lower than 10 cm (4 in) above the box's base as it needs to restrain chicks when the flap is open.

A glass window at the brooder's top helps with the monitoring of chicks.

Where there is no risk of vermin getting at the chicks, a complete covering to the brooder is not necessary.

To provide warmth for the chicks, suspend an unshaded, tungsten-filament light bulb 30–45 cm (12–18 in) above the brooder's base. Depending on the size of the brooder, a 100-watt or a 150-watt lamp is sufficient to raise the temperature to 35°C (95°F). Check this temperature with a thermometer at 7.5–10 cm (3–4 in) above the brooder's base. If the brooder remains cold, use a bulb with a higher wattage.

Another way to produce the correct temperature is to raise or lower the light bulb. A further method to adjust the temperature is to use a 'power reducer'.

After a few days – and when the chicks have settled down and are established in the brooder – start to lower the temperature by 1–2 degrees each week. Repeatedly check the chicks for signs of being too hot: this is indicated when they mainly go to the outside edges of the brooder. Conversely, if chicks cluster in a group in the centre of the brooder, this shows that the temperature is too low.

A recommended diet for young chicks is detailed on pages 150–151. They also need fresh water. Place a shallow dish of water in the brooder and gently dip their beaks into it so that they acquire the idea of drinking (remember that these chicks will not have a mother hen present to show them how to eat and drink, so it is up to you to help them).

Continue housing the chicks in a brooder, keeping them warm, fed and watered. Also, slowly reduce the temperature until the chicks are fully feathered.

Once the chicks are able to live without the help of artificial warmth, they can either be put in a haybox brooder (if the weather outside is cold; see below) or given an enclosed run of their own outdoors (in summer).

If outdoors, do not allow them to mix with other chickens because of the risk of transfer of diseases to them. Additionally, ensure that the ground is not infected with parasites from older chickens.

A haybox brooder is an ideal home for young chicks.

Introducing young chicks to a broody hen

This is an alternative way to look after newly hatched chicks, but does need a hen that has been broody for at least two weeks. Some broody hens accept an 'instant' family, while others will not and it is difficult to predict what her attitude will be. If she is cooperative, it is an ideal way to raise chicks newly hatched from an incubator. The ideal surrogate mother for day-old chicks is a hen that is two or more years old. Breeds that make good mothers are indicated in the breeds section of this book (see pages 28–61). Place the prospective mum in a large, straw-lined box that will give her and the young chicks warmth and protection.

The introduction of chicks to a new mum is best undertaken after dark. Gently put them, one by one, behind her so that initially she is unable to see them. There is a chance she will reject them by kicking backwards. If this happens, as soon as possible, put the chicks in an artificial brooder (see page 158).

If good fortune is with you and she accepts them, an indication of her willingness to become a surrogate mum is her clucks in response to their cheeps. A large-breed hen will take up to 12 chicks, and further offers acceptance of them by spreading herself around them. Leave her alone for a few hours, then check she has not changed her mind.

Haybox brooder

If the chicks need to be kept under cover, place them in a haybox brooder until well established. Basically, this is a wooden enclosure with sleeping and run areas. The sleeping area usually has a wooden, insulated lid, while the run is covered by fine-mesh wire to prevent vermin getting at them. Ensure that water and food is available to them (see pages 150–151 for details of the required food for growing chicks).

PESTS, DISEASES AND PROBLEM HABITS

Troubleshooting

Keeping groups of similar animals in relatively confined areas invariably encourages the presence of pests, diseases and problem habits. Therefore, apart from thoroughly inspecting chickens before introducing them into your group, continued vigilance is essential throughout their lives.

Keeping chickens in a small, congested area creates many problems for them.

What are the types of problems?

There are probably many more problems that can arise with chickens than you could have considered possible. Their range is wide and includes diseases, internal and external parasites, problem habits and stress-related disorders.

• **Internal parasites**: These are usually types of worms which when inside a chicken create problems. Caecal Worms, Gape Worms, Gizzard Worms, Roundworms, Tapeworms, Thread Worms and Coccidiosis (small parasites that live in the guts of chickens) are the main internal parasites (see pages 168–171 for details).

• **External parasites**: These are better known than internal parasites as they infest other warm-blooded creatures as well as chickens. They include Fleas, Lice, Mites and Ticks. They suck blood, causing great distress as well as transmitting diseases throughout your flock (see pages 166–168 for details).

• **Diseases**: These are often highly contagious and a couple, such as Newcastle Disease (Fowl Pest) and Avian Influenza (Bird Flu), are notifiable diseases (see page 165). Other diseases are Avian Enterohepatitis, Fowl Cholera, Marek's Disease, and Roup. Diseases are more difficult to detect than external parasites and therefore greater vigilance is needed to ensure they are identified at an early stage and controlled (see pages 172–173 for details).

• **Problem habits**: These are relatively obvious and daily inspection of your flock will reveal birds that are showing problems. They include bumble foot, cloacitis, crop Impaction, egg eating, internal laying, cannibalism, cramp, egg binding, feather pecking, and prolapse.

Unless treated, these problems cause great distress to chickens (see pages 176–179 for details).

• **Stress-related disorders**: Chickens are usually quiet, delighting in an unchanging routine where food arrives at the same time each day and is delivered by the same person, whom they know. Any change from their normal routine can upset them and cause problems such as diarrhoea, laboured and irregular breathing, and radical changes in their normal pattern of behaviour. Many of these problems can be easily prevented (see pages 174–175 for details).

An inquisitive face, bright eyes and attractive comb create the impression of a healthy chicken.

Signs of good health in chickens

Whenever you feed and water your chickens, or collect eggs, glance from one bird to another to compare them. Their colours and styles of plumage may differ, but overall you will be able to detect changes. Here are the main visual signs of good health:

★ **Bodyweight**: This should be correct for the breed and age of the chicken. See page 28–61 for the range of weights for adult birds (male and female).

★ **Breathing**: An even intake and expelling of air. There should be no 'rattling', and certainly without discharge.

★ **Combs**: These vary from one breed to another and should comply with the nature of the breed. See page 20 for details of combs.

★ **Droppings**: Firm and dark, with a white tip. There should be no sign of diarrhoea (see page 175).

★ **Eyes**: Must be bright and clear, with an eagerness to search for food.

★ **Feathers**: Should be smooth and shiny.

★ **Gait**: Upright and alert, not leaning and straining to stand upright.

★ **Legs**: Clean and strong and able to support the chicken, when standing still as well as walking.

★ **Smell**: This might appear to be a strange thing to check upon, but a bird in good health should emit a 'warm and pleasant' aroma.

★ **Tail**: Should be carried correctly for the type of breed.

Behavioural assessments of good health

Chickens are usually active during daylight hours and the way they move and react to each other reveals good health.

Here are the main behavioural signs to look out for:

- **Calmness**: A restful and unexcitable nature is desirable.

- **Feeding and drinking**: This should be normal for the time of day.

- **Laying**: Hens of the correct age (see pages 104–105) will be regularly laying eggs. If a hen stops producing them, this is a sign she needs to be inspected. It is not always a problem and may just be part of the hen's egg-laying cycle (see page 148).

- **Moving around as normal**: Usually, this is with an inquisitive and enquiring nature – and invariably throughout the day. If a hen stops being active, check her for problems.

- **Perching steadily and with confidence**: Part of this steadiness results from the correct size and height of the perch (see page 75 for details).

- **Preening**: Regular preening indicates that the bird is healthy and takes a pride in itself.

- **Putting on weight**: This will be most noticeable in pullets (see page 104).

- **Quiet, but not silent**: A chicken should exude a confident stillness and not be clucking and squawking and disturbing other birds.

- **Sparring or mock fighting**: This is often apparent in young birds, but should not be exceptionally aggressive.

- **Sunbathing or dust bathing**: These are ways in which chickens rid themselves of parasites.

Preventing the onset of pests and diseases

Regularly inspecting your chickens for signs of good health is essential, but there are other ways to keep your chickens happy and free from pests and diseases. Here are a few:

- **Clean living areas**: Chickens must be housed in clean poultry houses. Additionally, scratching runs need to be kept clean and, especially, free of droppings from wild birds which may carry diseases.

- **Clean food storage**: Stored food for your chickens must be kept in clean, sealable bins in secure sheds. Always label the container to ensure your chickens get the correct food (young birds need different food from adults).

- **Clean water**: Dirty, germ-infected water soon spreads diseases among chickens. Daily check and replenish drinking water; if the container is an open bowl and contaminated with faeces or decaying organic material, remove them and scrub it before refilling with fresh, clean water.

- **Balanced diet**: Ensure your birds are given the correct diet for their age (see page 87).

- **Kitchen scraps**: Although chickens like to eat leftover kitchen scraps, first check they are not decaying and have not become contaminated with chemicals (from kitchen cleaners, for example). Additionally, do not accept scraps of food from neighbours as you will not know where they originated.

- **Congested poultry houses and runs**: Do not cram your chickens together as this may cause stress, squabbling and fighting. Stress soon debilitates a chicken and makes it more likely to suffer from diseases and pests. The ideal number of chickens for a chicken house or ark is given on page 72.

- **Early problem detection**: If you notice a chicken that appears to be ill, you should immediately isolate it and seek veterinary advice.

See pages 162–179 for a range of problems, and how best to deal with them.

- **Wild birds**: Veterinary experts confirm that diseases can be transmitted from wild birds to domestic chickens.

Often, poultry keepers like to ensure wild birds have access to fresh water and therefore put down water containers especially for them – but well away from their chickens.

A chicken's temperature

Similarly to other warm-blooded creatures, the health of a chicken can be partly judged by its body temperature. A chicken's normal temperature is 40–41.5°C (104–107°F). However, this may fluctuate in response to the environment, whether hot or cold.

During very hot summers a chicken's temperature may exceed this range, but if it reaches 45–47°C (113–117°F) a chicken must lose body temperature quickly if it is not to suffer. High temperatures cause hens to stop laying and may result in death.

Fortunately, chickens are able to lose body temperature if it becomes excessive: their respiratory system, together with the ability to pant, in which they use the hot weather to evaporate water from their throats, enables their body temperature to be decreased.

To prevent chickens becoming overheated, ensure that the chicken house is well ventilated and with a roof that reflects heat. Conversely, it also needs to be insulated against low winter temperatures.

The scratching run needs to be kept cool and this is best achieved by positioning it under a deciduous tree. However, such trees may encourage wild birds to mass in its branches and defecate into the run.

Judging and assessing a chicken's temperature

If a chicken appears to be lethargic or panting, this is an indication that it is overheating and its temperature needs to be taken. Additionally, your chickens may show symptoms of illness caused by pests and diseases – these are detailed on pages 166–173.

Nowadays, there are digital thermometers and these devices make reading a chicken's temperature much easier, but the thermometer still needs to be inserted into a chicken's back passage.

When taking the temperatures of a group of chickens, thoroughly wipe and clean the thermometer between each 'patient' to reduce any risk of a disease spreading.

Notifiable diseases

There are a few diseases of chickens which, by law, have to be notified to official government organizations as soon as they are seen. These are:

★ **Avian Influenza (Bird Flu)** – see page 172 for details.

★ **Newcastle Disease (Fowl Pest)** – see page 172 for details.

At the same time, birds suspected of having one of these problems must be isolated in separate pens to keep them away from other chickens as well as wild birds.

Diseases that must be notified to government organizations vary from one country to another; therefore you must check with your local animal health and welfare agencies for details of them.

Parasitic pests

These live on the inside or outside of chickens. They cause physical debilitation, extreme irritation and loss of egg production. They may also result in a decrease in bodyweight. Once established, they are difficult to eradicate, and therefore regularly inspecting your chickens is essential.

External parasites

These are better known than internal parasites and infest chickens as well as other warm-blooded creatures. They include Fleas, Lice, Mites and Ticks.

When selecting a spray or dust to combat them, always check that it is suitable for use on chickens. If you are not sure about its use, ask a vet for advice.

Fleas

Few external parasites are better known than fleas, which are notorious when infesting dogs and cats.

- **Description**: Small, about 2.5 mm (¹/₁₀ in) long but, because of their enlarged rear legs, able to leap distances disproportionate to their size. There are several species that infest poultry.

Their mouthparts are enlarged and adapted to piercing skin and sucking blood. In this way they transmit diseases from one chicken to another. They cluster in groups on the skin and are often found in difficult-to-see places.

- **Life cycle**: Although infesting chickens, they spend most of their lives in bedding, which becomes a residual source of infection. They live for several weeks on a host chicken and may survive up to a year if able to return at regular intervals to suck blood. Female fleas lay eggs which hatch in the bedding.

- **Symptoms of attack**: The sucking of blood causes the victim to be restless and itchy.

- **Treatment**: Look for clusters of fleas, especially in areas where a chicken may be pecking in an attempt to remove the irritation. As soon as fleas are seen, dust them with a flea powder. Then remove all chickens to a clean and separate enclosure; take away and burn all bedding (such as straw and wood shavings).

Thoroughly wash and scrub the building, including nesting boxes. Spray the building with an insecticide and allow several days before refilling with clean bedding. Return the chickens, but inspect them every week to check if fleas are re-established.

Mites

These eight-legged creatures are more pernicious and devastating to chickens than fleas and even more persistent than lice. They are extremely difficult to eradicate.

Mite

- **Description**: They are about 1 mm (¹/₂₅ in) long and with a hard body. There are many different types of mite but the Red Mite (also known as the Chicken Mite) is the one most common in chickens.

Red Mites are grey in colour until they suck blood from their host, when they become red. They live and lay eggs in cracks in chicken houses, especially in nesting boxes and where chickens roost. They can survive for up to six months off a chicken.

- **Life cycle**: They usually feed on chickens at night and return to cracks and crannies in a poultry house during the day. However, if an infestation is severe they remain on birds during the day.

The reproductive rate of Red Mites is prodigious, with infestations increasing rapidly during hot, dry summers.

- **Symptoms of attack**: Mites cause immense irritation and, in extreme infestations, result in death, especially to young chicks. Sometimes, they invade the roof of a bird's mouth, resulting in anaemia.

Inspect the birds at night, when mites are most active. With the aid of a flashlight they often can be seen crawling on the birds.

- **Treatment**: Remove all chickens from their enclosure and place them in a clean shed. Spray them with a proprietary chemical control recommended for use on chickens.

A recent pesticide-free treatment contains fossilized remains of microscopic shells, called diatoms, which have microscopic razor-sharp edges that cut through the insect's protective covering. It has the advantage of being a totally organic method of control.

Remove and burn all bedding from within the infected poultry house. Water from a high-pressure jet will dislodge them; when dry, spray or dust the entire area with an insecticide.

Where mites are persistent and deeply embedded in cracks in timber, the use of a blow-lamp has earlier been recommended – but take care not to set the building on fire!

Ticks

There are two main types of tick – hard and soft. The hard type is mostly found on free-ranging chickens in temperate and tropical regions. The soft type is mainly seen in the southern states of North America, where it is locally known as Adobe Tick, Blue Bug, Chicken Tick and, chiefly, Fowl Tick.

Description: They are flattish, leathery, oval-shaped and resemble large mites, about 12 mm ($\frac{1}{2}$ in) long and with eight legs. They are tough and resilient, often persisting for four years between feedings on poultry. Initially they are reddish-brown or tan-coloured, but after feeding turn bluish.

Tick

- **Life cycle**: Like many other parasitic pests, they hide in crevices and cracks in poultry houses during the day and infest chickens and other poultry at night. Females lay several batches of eggs, each containing about 150 eggs which hatch approximately ten days later. In cool weather, their development is retarded and it may take up to three months for them to hatch.

- **Symptoms of attack**: They feed for 15–30 minutes on a chicken, leaving red spots on the bird's skin. Chickens often detect the presence of ticks and become restless when roosting.

The ticks attach themselves to bare skin under wings or around the head, sucking blood and causing anaemia. Egg production usually decreases during times of infestation.

- **Treatment**: As soon as they are seen on your chickens, remove the birds to a clean poultry house. Treat the birds with a recommended spray or dust. Then, use a power jet to dislodge the ticks from cracks in the building's structure and, when dry, dust with an insecticide.

Throughout the following weeks, regularly check that the ticks have not re-established themselves.

Lice

Even the word lice engenders thoughts of creepiness and filth. There are claimed to be more than 40 different species of them.

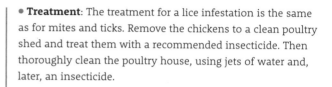

Louse

• **Description**: Poultry lice are the species most often seen on chickens. They are usually about 3 mm (⅛ in) long and range in colour from off-white to brown and dark grey. They have hard, scaly bodies and three pairs of jointed legs.

• **Life cycle**: A louse lives for several months and can go through its entire life cycle on a chicken's body. Indeed, it can only survive for a week – or less – when off a chicken's body. Female lice lay eggs (called 'nits') on a chicken's feathers. She ensures they remain in position by sticking them down with a glue-like material.

These eggs hatch within four to seven days and the young lice (known as 'nymphs') resemble very small versions of adult lice. Initially they are transparent, slowly passing through several moults until they look like parent lice.

When a nymph matures it mates on the bird and a female louse may lay as many as 300 nits in her lifetime. It therefore does not take very long for a single louse to become a major infestation – and a problem!

Lice spread from bird to bird by contact. Additionally, wild birds may pass them to your chickens.

• **Symptoms of attack**: Infestation is worse in autumn and winter; chickens become restless, pecking at themselves and checking for the presence of lice. By parting their feathers, you can often see the lice scurrying across the skin. They feed on blood, skin debris and at the roots of feathers. Depending on the species, some mainly infest the body, others the head.

• **Treatment**: The treatment for a lice infestation is the same as for mites and ticks. Remove the chickens to a clean poultry shed and treat them with a recommended insecticide. Then thoroughly clean the poultry house, using jets of water and, later, an insecticide.

Internal parasites

These are usually types of worms which, when inside a chicken, create problems. Caecal Worms, Gape Worms, Gizzard Worms, Roundworms, Tapeworms, Thread Worms and Coccidiosis (small parasites that live in the guts of chickens) are the main internal parasites.

Caecal Worms

Sometimes known as Cecal Worms, they are one of the most common parasitic worms in chickens. Some breeds of chicken, such as the Leghorn, are more susceptible to this parasite than heavier ones like Wyandottes and Rhode Island Reds.

Description: There are several species of this worm, ranging in colour from yellow-white to white and up to 18 mm (¾ in) long. They are present in the caeca (blind gut) of poultry and usually more of a problem in turkeys than chickens. This is because Caecal Worms carry the organism that causes Blackhead Disease (also known as Avian Enterohepatitis) in turkeys. Chickens that are free-ranged with turkeys are especially at risk from this problem.

• **Life cycle**: Adult worms lay eggs. Later, depending on the temperature but usually within 2–4 weeks (longer in cool climates), the eggs hatch in the soil and mature into worms.

• **Symptoms of attack**: Severe infestations cause inflammation of the caeca, which results in pale birds with a drooping habit. Infected birds also tend to huddle together. Additionally, they lose their appetites and drink less water; this is

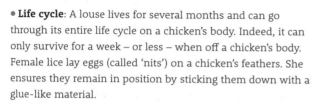

Caecal Worm

often accompanied by diarrhoea and leads to emaciation, dehydration and a marked decrease in egg production.

● **Treatment**: Consult a vet immediately birds are seen to be in distress and treat with a recommended worming medicine.

Coccidiosis

This is the most common protozoal disease of poultry. There are several species of Coccidia and they live in cells lining the gut, causing irritation and interfering with the absorption of nutrients. Eventually, the gut is destroyed.

● **Description**: These are single-celled creatures, too small to be seen without the aid of a microscope.

● **Life cycle**: This is short but complex. A chicken when pecking at the ground may ingest an infected cyst which, when in the bird's gut, releases sporozoites which lodge in the bird's intestinal wall. They begin to reproduce and within four to six days several more generations are present; it is a very rapid cycle and birds soon become debilitated. Infected chickens spread this problem through their droppings. Additionally, infected cysts can live in the soil for up to 18 months, depending on the temperature and the soil's moisture content. Warm weather and damp soil create ideal conditions for their survival. However, they do not survive temperatures below freezing and above 55°C (130°F). Because reinfection often results from birds pecking at contaminated soil, control is difficult. Also, infected as well as recovering birds shed cysts in their droppings.

● **Symptoms of attack**: The problem develops into watery diarrhoea which may be stained by blood. Birds lose weight, become distinctively sick and ruffled-looking. Egg laying rapidly decreases and, eventually, stops.

● **Treatment**: Consult a vet immediately symptoms are noticed and treat with a recommended medicine. Remove birds to a clean, uncontaminated area and keep it clear of faeces. If birds are in an ark, move it to a clean piece of ground. Free-ranging chickens are more at risk of contamination – and reinfection – than birds that are kept in an enclosure.

Gape Worms

Often known as Red Worms and Forked Worms, they infest many types of poultry, especially chickens.

● **Description**: These worms are 6–18 mm ($\frac{1}{4}$–$\frac{3}{4}$ in) long, red and Y-shaped, and can be readily seen.

● **Life cycle**: An infected chicken coughs up worm eggs, then swallows them. These are then expelled in its droppings. Eggs may survive in the soil for up to four years; they are also part of a cycle of infection involving worms, slugs and snails. Therefore, once chickens and the soil are infected it is difficult to free both birds and the soil from this debilitating trouble.

● **Symptoms of attack**: If you see a bird repeatedly gape, its mouth opening but not making any noise, this is Gape. Birds also shake their heads and have difficulty in breathing. Additionally, infected birds show a loss of appetite. Young birds are especially badly affected. Repeated yawning, gasping, sneezing, coughing, choking, grunting and closure of eyes are other symptoms. Gape Worms bury their heads in the lining of a bird's windpipe and other parts of its respiratory system, resulting in a 'gaping for air'.

● **Treatment**: Consult a vet, isolate the infected chicken and treat with a recommended worming medicine.

Rapid treatment
Gape Worms are highly debilitating to a chicken and the sight of a bird making involuntary gapes can be frightening to poultry keepers. Therefore, as soon as symptoms appear, gain advice from a vet. Do not wait for the problem to resolve itself – it won't.

Gizzard Worms

These pernicious and tenacious worms are usually a problem with geese, although chickens may also become infected when grazing on the same land.

● **Description**: Fine, thread-like worms, reddish and 12–25 mm ($^1/_2$–1 in) long.

● **Life cycle**: This parasitic worm uses grasshoppers and beetles as intermediate hosts in its life cycle. Spring and early summer are the times when your chickens are most at risk of infection.

● **Symptoms of attack**: Fine, cotton-like worms become attached to the inner linings of gizzards. Infected birds stagger, lose weight and become weak and ill. Older birds tend to survive an infestation, but young ones are more susceptible and require immediate attention.

● **Treatment**: Do not allow chickens to free-range on infected land and especially in the company of geese. Land surrounding a communal pond can be a serious source of infection.

Early treatment is essential for infected birds and this involves contacting a vet and adding a recommended worming medicine to the birds' drinking water.

Roundworms

Most chickens and other poultry can survive slight infestations of Roundworms, but it is when they reach high levels that problems arise.

● **Description**: These worms are readily identifiable – up to 10 cm (4 in) long, thick, oval and yellowish-white. They can often be seen in, and alongside, a bird's droppings. Adult roundworms infest a chicken's small intestine.

It is essential to keep chickens free from roundworms.

● **Life cycle**: Roundworms soon multiply and a female lays up to 5000 eggs. They are able to live in soil for a year or more and therefore are a major source of reinfection.

● **Symptoms of attack**: Listlessness, poor growth and development, loss of condition and diarrhoea are the most common signs of infection.

● **Treatment**: If young birds are not treated rapidly, they may just waste away. Consult a vet and either give pills to your chickens or add a worming medicine to their drinking water. Good hygiene is essential to prevent reinfection; keep feeding bowls and water containers clean and free from faeces. Additionally, keep chickens off contaminated land as earthworms and grasshoppers transmit roundworms.

Tapeworms

These are classified as 'Flatworms', and it is estimated that up to 50 per cent of all chickens in small flocks are infested with them. Some breeds, such as Leghorns, have slight resistance to these worms.

● **Description**: Several species of Tapeworm infest chickens and some are near-microscopic, while others are up to 25 cm (10 in) long. They lodge in the intestinal tract, with their heads attached to it by four pairs of suckers. Each species of Tapeworm infests a different part of the intestine.

● **Life cycle**: All Tapeworms have an intermediate host and this may be a beetle, ant, earthworm, slug or snail. Flies are another host, especially when birds are kept in cages. But mostly it is earthworms, slugs and snails that are in their life cycles, especially for free-ranging birds.

A Tapeworm's body is formed of individual segments and one or more of them break away each day. A chicken starts to shed segments within two or three weeks of being infected, usually through eating an intermediate host. It is also possible for a chicken to peck at and eat Tapeworm segments on the ground; each of these contains several hundred eggs.

- **Symptoms of attack**: Lethargy, difficulty in breathing and loss or weight are the main symptoms of infection. Segments (which often look like pieces of rice) may also become attached to the vent (backside) area of a chicken; they can also be seen in faeces.

- **Treatment**: Consult with a vet at the earliest opportunity as Tapeworms can be difficult to eradicate. Free-ranging birds are more likely to become infected than those kept in runs.

Thread Worms

Also known as Capillary Worms, these are minute with several species infesting different parts of a bird's internal organs. This variability results in many different symptoms, but mostly they infect the upper part of the intestine.

- **Symptoms of attack**: Diarrhoea, with severe infestations ultimately causing death. It is difficult to see these worms in a chicken's faeces and therefore identification of the problem can be difficult. As the diarrhoea increases, a bird becomes more debilitated and distressed.

- **Treatment**: As soon as you suspect Thread Worms to be infesting your chickens, you should consult with a vet and initiate treatment.

Free-ranging chickens usually have a healthier life than those tightly confined in one area.

Diseases of chickens

Diseases can cause major problems for chickens and it is essential to identify them at an early stage in their development. The characteristics of some diseases are initially similar to one another and therefore it is essential to call upon the experience of a vet if you are not sure about the problem.

Newcastle Disease (Fowl Pest)

This is a major killer of chickens, especially young birds in large, commercial flocks. It gains its name from the city of Newcastle-upon-Tyne, one of the first places where the disease was studied. In many countries this disease is known as Fowl Pest. However, the term Fowl Pest can be confusing as it also refers to Fowl Plague. Therefore, for clarity it is best to refer to this disease solely as Newcastle Disease.

This is a disease that attacks many species of birds in the wild and for that reason is rapidly spread from one country to another, infecting native birds and especially those kept in large commercial flocks.

There are several degrees of severity with this pernicious and rapidly spreading disease, with some forms being mild enough to be contained by injections. Usually, however, it is severe, with all infected birds dying.

- **Symptoms**: It can affect chickens suddenly and without warning. Infected chickens sneeze, gasp for air, become listless, have muscular tremors, wing or leg paralysis, twisted necks and, sometimes, watery and greenish diarrhoea. As can be expected, there is a drop in egg production, accompanied by misshapen or soft-shelled eggs.

- **Treatment**: Once your chickens are infected there is no cure and they must, without hesitation, be isolated. Immediately contact your local Animal Health office. Additionally, until you are given permission no one can enter of leave the premises. It is a notifiable disease (see page 165) and must be taken seriously to prevent its spread.

Avian Influenza (Bird Flu)

This is another notifiable disease (see page 165). It is a viral problem and chickens and turkeys are especially at risk, although all poultry can be become infected. It is often spread by wild birds.

There are several different forms, ranging from mild to fatal.

- **Symptoms**: These vary, but relate to respiration and encompass coughing and sneezing. Additionally, there is a loss of appetite, diarrhoea, droopiness and depression, sudden decrease in egg production, blood-tinged discharge from the nostrils, distress when breathing, swelling of the head, eyelids, comb wattles and hocks. As if this is not enough, an infected chicken also reveals lack of coordination when walking and standing. Sometimes, there is sudden death in apparently healthy birds.

- **Treatment**: At the first sign of infection, contact your local Animal Health office. Additionally, isolate your flock and ensure that until you are given permission no one leaves or enters the premises.

Marek's Disease

Also known as Fowl Paralysis Virus, it is a highly contagious viral disease and usually results in the deaths of chickens unless they have been vaccinated when day-old chicks (immunity develops within two weeks from the day of injection). A single vaccination gives a chick life-long immunity, but is not foolproof and about 5 per cent of vaccinated chicks will get the disease. There is no known cure once a bird has become infected.

This disease kills more chickens than any other infection in North America, where it is widespread. Chickens which have been under stress and weakened from overcrowding or being moved are especially at risk of succumbing to the virus.

• **Symptoms**: Chicks which have not been vaccinated are usually 4–8 weeks old at the onset of symptoms, which include paralysis of the wings, neck and legs. This is accompanied by loss of weight and a distinctive 'straddling' stance, with one leg held forward and the other back. Paralysis is progressive. Vision is also impaired, with grey irises and irregular pupils.

• **Treatment**: Once a chick or chicken is infected, treatment is not possible. As a preventative, chicks can be vaccinated (see above).

Avian Enterohepatitis

Also known as Infectious Enterohepatitis, it is mostly seen in turkeys and known as Blackhead. Chickens, quails, guinea fowl, grouse and partridges are also affected. It is caused by a protozoan organism which is transmitted either in the droppings of an infected bird or in the eggs of the Caecal Worm (see page 168). These eggs survive in the soil for several months and, later, may be ingested by a bird.

• **Symptoms**: The name Blackhead results from infected birds developing dark and discoloured heads. Drooping wings, drowsiness, stilted gait, closed eyes and walking with head downwards are other symptoms.

• **Treatment**: Keep turkeys on land separate from chickens. Also keep the bedding clean and gain advice from a vet.

Fowl Cholera

A pernicious and highly contagious disease that affects chickens, turkeys, ducks and geese. Unless treatment is given rapidly to an infected fowl it can be fatal. A hen may appear perfectly healthy one day, but be found dead during the following one.

• **Symptoms**: Ruffled feathers, greenish-white diarrhoea, severe thirst, coughing, swollen joints, loss of appetite and a nasal discharge are just a few of the unpleasant symptoms.

• **Treatment**: Immediately these symptoms are seen, contact a vet. Infected birds must be killed and disposed of in a legal manner (see page 137). There are preventative drugs and these are added to the drinking water.

Roup

If a common cold is neglected, it may turn into Roup, which reveals itself in two highly contagious forms – Wet Roup and Dry Roup. In general, Roup is associated with Vitamin A deficiency and sometimes known as Nutritional Roup.

• **Symptoms**: Wet Roup produces an offensive discharge from a chicken's respiratory tubes. Dry Roup, however, is similar but with comparatively little discharge. Other symptoms include poor growth, drowsiness and eyelids that are often stuck shut by a thick and sticky exudation.

• **Treatment**: Add Vitamin A to the drinking water. Where the disease is seen early in its development, quickly isolate the infected chicken. It is a distressing problem for chickens. Take advice from a vet.

Stress-related disorders

These can be just as debilitating to chickens as an infestation of pests and the onset of a disease. Chickens invariably have disagreements with each other, with one bird usually becoming dominant in the pecking order (see page 16), but usually they are at their happiest and contented when in a group of three to six.

Stress management

Some diseases are so virulent that they invade chickens whether or not they are under stress. Others are milder and only enter a chicken and cause problems when it is under stress. It is therefore essential to keep chickens in an unstressed and calm situation at all times.

Occasionally, stress cannot be avoided and is part of a chicken's everyday life. Hatching, moulting and reaching maturity cause stress, but these are natural happenings and cannot be avoided. Others result from environmental factors and include low and high temperatures and excessive humidity – and these are the ones that can be avoided.

Hens are usually more easily stressed than cock-birds, although when they are in a group they offer each other reassurance that 'all is well'.

Preventing and avoiding stress

Ways of doing this include:

● **Being calm when handling chickens** and moving carefully among them. Sudden and unexpected moves will cause alarm.

● **Not overcrowding** your chickens (see page 72).

● **Providing clean straw and bedding** for them.

● **Ensuring the enclosure is free** from external parasites (see pages 166–168) and well secured against vermin.

● **Regularly providing food and clean water** at the times the chickens have come to expect. Where possible, do not change the source of water given to them as different-tasting water will confuse them.

● **Where possible, ensuring that noise,** such as that from motor vehicles, fireworks and barking dogs, cannot alarm them. Also, avoid having smoke from a neighbour's barbecue blowing over them.

Chickens are gregarious and delight in attention. Take care, however, that young children are safe among them.

- **Handling and talking** to all of your chickens in a gentle manner. Those that are ignored and treated roughly are more likely to become infected with diseases.

- **When engaging a 'chicken sitter'** to look after your birds when you are away on holiday, asking the person to visit your chickens several times during the previous week so they can get to know each other.

- **Never making rapid changes** to their living quarters and pens, it will confuse them.

Signs of stress behaviour

Although the signs of stress vary, they can be grouped under three headings. These are:

- **Laboured and irregular breathing**: This is usually brought about by panic; it can also be through high temperatures and overcrowding. It may also result from a respiratory problem (advice from a vet may be needed).

- **Diarrhoea**: The mildest form of this is when a chicken poops on you after being roughly caught and held. If, however, the poop is more and radical, something serious may be wrong and advice from a vet is needed.

- **Changes in normal behaviour and daily activities**: These are many and varied, with chicken behavioural experts offering many categories into which they can be grouped:

- **frequency of eating and drinking** is reduced, even though a chicken may need water.

- **flinching and shying away** from sudden movement.

- **a frightened chicken will shake its head** vigorously from one side to the other.

- **stressed chickens move up and down** their 'runs', in a similar way to caged animals in zoos.

If your chickens appear unaffected by their surroundings and continue to eat and drink, they are probably 'happy'.

- **birds become 'flighty'**, moving in an unpredictable manner.

- **chickens lose interest** in keeping themselves clean and smart. They stop dust bathing, which is the normal method they use to keep their feathers clean and free from parasites.

- **birds become restless**, especially at night when roosting.

- **birds stop investigating** new happenings.

- **the pecking order becomes more noticeable**, with birds at the lower end being excessively dominated and chased away from feeders and water points.

Problem habits in chickens

Pests and diseases – as well as stress-related disorders – are not the only difficulties you may encounter when keeping chickens; there are also 'problem habits'. Most of these can be readily identified and home treatment quickly given to the bird. However, if you are in any doubt about a diagnosis, you should consult a vet at the earliest opportunity.

Bumble foot

This is one of the easiest problems to identify; the chicken, instead of walking evenly, hobbles and appears to have a bumbling gait.

Bumble foot

● **Symptoms and description**: It is caused by a small wound on a foot that does not heal properly. The outside of the wound appears to heal, but remains infected internally and results in a hard abscess on the ball of the foot. This becomes so painful that the hen cannot walk properly and therefore hobbles.

● **Treatment**: Use a sharp, sterilized scalpel or knife to cut open the raised area to enable pus to escape. It is necessary to squeeze out all pus and to ensure its hard, central core is removed. When the area is clean and free from infection, apply an antiseptic cream.

Place the bird in a clean, dry pen with fresh straw until the wound fully heals, usually within three to five days.

Wounds sometimes occur if a perch is too high and the bird lands awkwardly on a hard surface (see page 75 for the recommended height and size for a perch).

Feather and vent pecking

A problem usually associated with chickens kept close together in runs: free-rangers rarely succumb to this habit. If neglected and untreated, it can lead to cannibalism (see opposite page), especially if it initially results in bleeding.

★ **Symptoms and description**: Areas mainly pecked and damaged are the back, rump and tail, with flesh torn and sometimes bleeding.

★ **Treatment**: Remove seriously affected birds and place in isolation in a quiet pen. Also, put the aggressive bird in solitary confinement for a few days. If, however, an exceptionally aggressive bird continues to attack fellow birds, permanent isolation (if you have space and the facilities) may be the only answer.

Giving the birds extra food and hanging up green stuff usually takes attention away from pecking their brethren. If the problem continues, you may need to have their beaks trimmed (but first ask the advice of a vet).

Chickens displaying signs of feather pecking. Cannibalism at its worse results in bare and torn skin and feathers.

Cannibalism

As the name implies, this is when a chicken attempts to eat the flesh of another chicken. It can occur in all types of housing systems, including free-range flocks, and is not restricted to chickens. Ducks, turkeys, quails and pheasants are also known to have cannibalistic natures.

Light breeds, such as Leghorns and other Mediterranean types, are often more susceptible to attack than heavy breeds. However, birds with cannibalistic tendencies are known in all breeds.

• **Symptoms and description**: A bird pecks, tears and consumes the skin and tissue of another in its group, causing great distress and injury to the victim. It usually begins with feather pecking (see opposite page) and, although a chicken's front is often the first part to be pecked, toes, tail and the rear become attacked areas.

Primarily, it is initiated by chickens being overcrowded or not having sufficient to eat. It also occurs when the pecking order (see page 16) has not been established, or when a new bird is introduced into a group.

Changing their diet to a less tasty one can sometimes also trigger cannibalism.

• **Treatment**: Once established in a group, cannibalism is difficult to control and therefore preventing circumstances arising that initiate the problem is vital. Always give the birds sufficient space (see page 72 for details of the space each chicken needs) and regularly check that undersized birds are not being bullied. Check that the diet is correct and in sufficient quantity. Additionally, move vulnerable birds to another pen.

Crop impaction

Sometimes known as 'crop bound', it is a digestive problem causing a chicken great discomfort and stress.

● **Symptoms and description**: It usually occurs in mature chickens. Crops become packed with food that cannot be passed out in the normal way. The crop becomes greatly distended, with a sour smell.

It is often caused by long pieces of grass being chewed and twisted into a ball. Occasionally, a chicken may eat a piece of string that becomes consolidated in its crop.

Birds affected by crop impaction stop eating in their normal way. Eventually, they cease eating.

● **Treatment**: Give the chicken a drink of clean, slightly warm water. This distends the crop, enabling it to be massaged to ease the blockage. Do not expect the blockage, when treated in this way, to clear quickly. However, it is a method that does not create too much stress for the bird.

A more robust measure is to hold the hen upside down and gently squeeze out the blockage. When free of this problem, allow her to drink freely; place her in a quiet pen until fully recovered and do not give her food for 24 hours.

As preventative measures, keep grass short throughout summer and do not leave pieces of string where hens, who are naturally inquisitive, can peck at them.

Cramp

This is a common and painful problem in poultry, affecting ducks more than chickens. It arises through imperfect blood circulation and is aggravated by cold and damp conditions underfoot. It is also considered to be the result of fatigue, lack of exercise and faulty feeding.

● **Symptoms and description**: The bird uncontrollably and suddenly squats down, with its toes contracted in an awkward manner. Walking becomes impossible, with the bird being unable to maintain control and balance.

● **Treatment**: Move the hen to a dry pen so that her feet are not damp. A layer of fresh straw helps to keep feet warm; leave her in a quiet pen until she has fully recovered and can walk properly.

The problem is intensified by dietary deficiencies such as in calcium and vitamin D; where necessary, add these supplements to her food.

Egg eating

Usually, this problem arises by accident, when a hen lays an egg which breaks on hitting the floor or other hard surface.

● **Symptoms and description**: Hens are inquisitive and likely to cluster around a broken egg, which invariably they find tasty. Unfortunately, this sometimes leads them to breaking their own eggs and eating them (usually occurs in older hens).

Other causes are nesting boxes positioned in strong and direct light, which can disturb a hen's natural desire for privacy when laying eggs.

Also, hens occasionally lay eggs without shells, a result of lack of calcium in their diets. Other reasons are when hens are congested or bored with life.

● **Treatment**: Give your hens plenty of space and hang up green stuff to give them something to peck at to relieve tedium in their lives. Also, add calcium supplements to their diets.

Egg binding

An egg-laying disorder usually resulting from the oviduct being too small. The oviduct is the long tube where eggs form and which delivers them to the hen's rear, from where they are laid. An illustration showing the oviduct and cloaca is featured on page 98.

- **Symptoms and description**: At its simplest, this is a problem when a hen cannot lay an egg easily and properly. It causes the hen great distress.

Young pullets are especially likely to suffer egg binding when first starting to lay eggs. It may also be caused by an egg becoming broken within a hen and not slipping out easily.

- **Treatment**: This is not easy and great care is required to prevent the hen being harmed. Expose her vent area for a brief period to hot water-vapour from a boiling kettle. This softens and eases her pelvic bones. Additionally, give her olive oil and place her in a straw-lined box in a quiet area. After a few hours she is usually able to lay an egg. It may be slightly streaked with blood.

Usually, the problem of egg binding disappears within a couple of weeks, but if you are anxious about it consult a vet.

Incidentally, hens when first laying eggs may produce long and narrow ones, but this is natural and should not alarm you.

Prolapse

This problem is also known as 'down behind' and usually happens to hens after their second year of laying. It rarely happens to young hens.

- **Symptoms and description**: It results when the inner walls of the cloaca are pushed out from the bird's vent area. Overweight and fat hens are especially prone to this problem and it usually results from the attempted laying of an oversized egg. It can also happen through excessive straining in an attempt to lay an egg.

- **Treatment**: It is a problem that needs to be treated as soon as noticed as it causes great discomfort to a hen. If left untreated, it may lead to cannibalism in the rest of the group. For details of cannibalism see page 177.

Gently, but firmly, support the hen with her head downwards; use warm water to clean the vent area. Take care not to cause unnecessary discomfort or to alarm her. Cover the area with olive oil and use a finger to work it into the orifice; gently, try to push the tissue back into place. Then put the hen on her own in a warm, straw-filled box in a lightly shaded and quiet area; leave her there for about a week. Do not excessively feed her but give plenty of water during this period. If the problem persists – and it is a condition that is likely to recur – repeat the treatment.

Cloacitis

Also known as 'pasted vent' and 'vent gleet', it is an unpleasant discharge of a milky-white substance from the vent area. Usually, it is considered to be a 'problem habit', but in reality is a highly infectious and contagious disease. Do not breed chicks from eggs that have been laid by infected hens.

Isolate affected hens as soon as symptoms are identified and, preferably, call in a vet. Usually, this results in the chicken being killed.

Internal laying

This is when an egg takes the wrong channel within a hen. It is unable to be passed out by the hen and invariably becomes infected, leading to peritonitis. Little can be done to help the hen and therefore she is best put out of her misery at the earliest opportunity; do not hesitate to take action.

Glossary

Poultry-keeping terms differ from one country to another and many of them are explained in this extensive and detailed glossary. Some of these terms are mainly associated with commercial poultry-keeping enterprises; others are known to keepers of chickens in their back gardens and yards. Whatever their derivation and association, they provide a better understanding of poultry.

Addled egg
Fertile egg in which the embryo has died after initially starting to grow. It may also refer to an egg that has begun to go bad.

Aerobic
Refers to microbes that require oxygen to enable them to develop and multiply.

Air cell/pocket/sac/space
It occurs in the wider and larger end of an egg and between the inner and outer shell membranes, where it provides a chick with air before it breaks free from the shell. This air cell can be seen when an egg is held up to a light. The space increases as an egg ages and is an indication of its age.

Albumen
This is the so-called 'white' part of an egg and it is formed of four alternating thick and thin layers. It contains about 40 different proteins, in addition to water. The albumen surrounds and protects the yolk (the yellow part of an egg).

Alektorophobia
The fear of chickens.

American breeds
Loose term for breeds developed in America and revealing common characteristics such as yellow skin, non-feathered shanks and red earlobes.

American Standard of Perfection
A book published by the American Poultry Association describing each breed recognized by their organization.

Amino acids
The form in which proteins are absorbed into the bloodstream.

Anaerobic
Microbes not requiring oxygen to develop and multiply.

Ancona
Light, soft-feathered breed of chicken (see page 31).

Andalusian
Rare light chicken breed (see page 31).

Appenzeller Spitzhauben
Light breed of chicken (see page 32).

Araucana
Light, soft-feathered breed of chicken, known in North America as the South American Rumpless (see page 32).

Ark
Small, relatively low, portable chicken house with a ridged roof; suitable for a small number of birds. It can be easily moved from one position to another.

Autosexing
Term for breeds in which the sex (male or female) of day-old chicks can be determined. This enables female chicks to be kept and male ones removed.

Axial feather
Short wing feather present between primary and secondary feathers.

Baby chick
Newly hatched chick, before it has been fed and given water.

Bantam and Miniature breeds
These two groups of chickens are often confused with each other: whereas True Bantams do not have large counterparts, Miniatures do and are just smaller-sized versions of large breeds. Unfortunately, even these definitions can be confusing as some breeds are classified differently from one country to another. Both True Bantams and Miniatures eat less food than large fowls and are less expensive to keep.

Banty
Affectionate term for bantams.

Barnevelder
Heavy, soft-feathered breed of chicken (see page 33).

Barnyard chicken
Chicken of mixed and indeterminate parentage.

Barred Rock
Heavy breed of chicken (see page 33).

Barring
Alternating dark and light stripes across a feather. Some breeds of chickens, such as the Barred Plymouth Rock, have this distinctive characteristic (see pages 18–19 for a range of feathering).

Battery
Intensive method of keeping chickens, where they are kept close together in wire cages.

Beak
Hard, protruding part of a bird's mouth, formed of a lower and upper part. It is adaptable and able to peck and crush food.

Beard
Group of bunched feathers under the beaks of some breeds, such as Faverolles and Houdan. See also Bib.

Bedding
Materials used to form comfortable floor covering for chickens and other poultry. It includes wood shavings and straw.

Belgium
Breed of True Bantam (see page 58).

Bib
Another name for Beard (see above). Additionally, it may refer to white markings on specific breeds of duck.

Biddy
Specifically, it means a laying hen more than one year old, but generally it is used as an affectionate word for a hen.

Black Australorp
Heavy, soft-feathered breed of chicken (see page 34).

Black Rock
Breed of chicken (see page 34).

Black Silkie
Breed of chicken (see page 35).

Blade
The part of a male bird's comb that is to the rear of its base.

Blastodisc
On the surface of the yolk there is an area known as the germinal disc; in unfertilized eggs this is called the blastodisc, while in fertilized eggs it becomes the blastoderm and plays an essential role in the progression to the next generation.

Bleaching
Fading of colour from the beak, shanks and vent of a yellow-skinned hen.

Blood ring
Signifies an early embryonic death.

Blood spot
Blood in the yolk or white part of an egg. It is caused by a slight rupture of blood vessels during its formation. It can be detected by candling and is harmless to the eater.

Bloom
The moist, protective coating on a freshly laid egg. Another meaning is having an exhibition bird in peak condition.

Blow out
Damage to the vent area of a chicken, caused by laying an oversized egg.

Bluebelle
Hybrid chicken (see page 36).

Blue-laced Wyandotte
Heavy, soft-feathered breed of chicken (see page 36).

Boiler
Old hen, perhaps one earlier kept for its ability to lay eggs which has now radically diminished. Such chickens require thorough cooking to make them suitable for eating.

Booted
Having feathers on the toes and shanks. This characteristic is seen on Booted Bantams, Sultans, and Belgian Bearded d'Uccles.

Broiler
Young bird specially raised for its meat.

Brooder
Piece of equipment for the artificial rearing of young birds.

Brooding
Raising newly hatched chicks in a protected environment.

Broody
Describes a hen with a natural urge to sit on eggs to hatch them.

Buff Orpington
Heavy breed of chicken (see page 37).

Buff Sussex
Heavy breed of chicken (see page 37).

Bumble foot
Problem habit with chickens (see page 176).

Caecal Worms
Internal parasite of chickens (see page 168).

Candling
A method to determine the interior quality of an egg by examining it against a light (originally a candle).

Cannibalism
A habit of some chickens when in a flock to attack and eat each other. Sometimes, it results from having too many birds in a small area (see page 177).

Cap
Comb (see Comb). It may also be the back part of a fowl's skull.

Cape
Feathers under and at the base of the neck hackle (see Hackle) between the shoulders.

Capon
A male chicken that has been feminized, either by surgery or chemically, to increase its speed of growth and improve its eating quality (see pages 132–133).

Carriage
The style and attitude of a bird; the way an individual bird carries itself as it moves.

Chalaza (plural: chalazae)
There are two of these within an egg and positioned at opposite ends of the yolk. They anchor the yolk to the inner ends of an egg, keeping it safe and in place and ensuring it does not become damaged if the egg is moved suddenly and roughly.

Checks
American term for cracked eggs.

Chick
Young chicken from the time it hatches to when it feathers out.

China eggs
Also known as 'crock eggs' and 'pot eggs'; they are dummy eggs for placing under a hen in order to encourage her to lay eggs.

Clean-legged
Describes a chicken without feathers on its shanks.

Cloaca
Part of the rear of a chicken's food canal, from which waste material is ejected, as well as eggs.

Cloacitis
Problem habit in chickens (see page 179).

Clubbed down
Abnormality in which down (the soft, first feathering of a young bird) appears in small, beaded nodules.

Clutch
The number of eggs a chicken (when in the wild) would lay before sitting on them to encourage hatching. Nowadays, this term usually refers to the number of eggs a poultry keeper allows a hen to sit on for hatching; it is influenced by the size of the hen.

Coccidiosis
Internal parasite of chickens (see page 169).

Cochin
Heavy, soft-feathered breed of chicken (see page 39).

Cock
Male bird aged 12 months or more that has completed one breeding season.

Cock-bird
Male bird aged 12 months or more.

Cockerel
Male bird under 12 months of age and in its first breeding season.

Comb
Fleshy prominence at the top of the head of a fowl, especially on a cock-bird.

Concentrates
Additives that provide a chicken with ready-make proteins, vitamins and minerals when mixed with their food.

Coop
Cage or house in which chickens live.

Cramp
Problem habit in chickens (see page 178).

Cream Legbar
Heavy breed of chicken (see page 39).

Crest
A crown or tuft of feathers on the heads of many breeds of chicken. It is sometimes known as the top-knot and found on breeds such as Polish, Houdan, Silkie and Sultans.

Croad Langshan
Heavy, soft-feathered breed of chicken (see page 40).

Crock eggs
See China eggs.

Crop
Enlargement in the gullet where food is stored and prepared for digestion (see page 83).

Crop bound
Clogging in a crop by bits of grass or, occasionally, straw and hay. Clogging may also be caused by pieces of string.

Crop impaction
Problem habit in chickens (see page 178).

Cross-bred
The first generation from crossing two different breeds or varieties. See also Hybrid vigour.

Cuckoo Maran
Heavy, soft-feathered breed of chicken (see page 40).

Cull
Removal and killing of surplus, ageing or infirm birds or chicks.

Cuticle
Thin layer of organic material covering the shell of an egg; it has a bloom that helps to seal the shell against rapid moisture loss. Additionally, it keeps out fine dust and bacteria.

Dam
Occasionally used to describe a mother hen.

De-beaking
Cutting off part of the beak to prevent feather pecking and cannibalism.

Down
The soft, first feathering on a young bird.

DPK
Sometimes used to refer to a domestic poultry keeper.

DPL
Dried poultry litter, a mixture of manure and floor material from floor-housed birds.

DPM
Dried poultry manure.

Droppings
Chicken manure.

Dual-purpose breed
Breed of chicken that both lays good eggs and is excellent as a table bird (for killing and eating). See page 134 for a list of suitable breeds.

Dubbed
Sometimes known as Dub, this is when the comb, wattles and earlobes are trimmed on an exhibition bird. This somewhat barbaric practice originated when fighting cock-birds were trimmed in this way to decrease the surface area an opponent bird could attack.

Dust bath
A box or depression in the ground filled with dry earth, sand or sawdust in which hens can frolic to keep themselves cool and clean.

Dusting or dust bathing
The act, on the part of a chicken, of thrashing and frolicking in dirt to clean its feathers and to discourage the presence of body parasites.

Dutch
Breed of true bantam (see page 58).

Ear tufts/whiskers
Feathers that grow directly from the earlobe region. Some breeds have ear tufts that curve and form a circle; these are a characteristic of the Araucana breed (see page 32).

Egg binding
Blockage of the oviduct by an extra-

large egg or an egg that has broken internally (see page 178).

Egg eating
Habit in some hens of eating their own eggs (see page 178).

Egg membranes
There are inner and outer membranes within an egg; these are positioned between the inner part of the shell and the albumen. They are transparent and form a defence against bacterial invasion.

Egg tooth
Horny protuberance on a hatching chick's beak which helps it to break out of its shell. Later, this egg tooth falls off.

Embryo
Young organism in the early stages of development. In a chicken, this is before hatching from an egg.

Endoparasites
Internal parasites, such as worms (see page 162).

Enzyme
Agent that helps in the breaking down of food in the gut.

Evisceration
The removal of internal parts of a chicken (and other poultry) when preparing it for eating.

Face
The skin around and below the eyes.

The colour is usually white, red or purple.

Faverolles
Heavy, soft-feathered breed of chicken (see page 41).

FCR
Refers to the conversion of food eaten to the resulting bodyweight. For example, 2 kg of food resulting in 1 kg of live weight is FCR 2:1 (or FCR 2).

Feather-legged
Having feathers growing down the shanks (a section of a leg and just above the foot).

Feather pecking
Habit of some hens of pecking feathers from other birds (see page 176).

Feathering out
The stage when a chick loses its initial, fuzzy coat and grows its first feathers. Usually, this happens 6–12 weeks after hatching, depending on the breed.

Fertile
Having the ability to produce a chick.

Fleas
External parasites of chickens and other warm-blooded creatures (see page 166).

Flight feathers
Primary feathers on a wing; sometimes denotes primaries and secondaries.

Flights
Main flight feathers (also known as primaries).

Flock
Group of chickens living together.

Flogging the hen
Earlier-used term for when a cock-bird mounts a hen to mate with her.

Fluff
Soft, down-like feathers running from the abdomen to the rear of the thighs.

Fowl
Collective term for chickens, ducks and geese.

Fowl Pest
Notifiable disease (see pages 165 and 172).

Free-range
Method of keeping chickens in which they are allowed to wander freely and to pick up insects and wild food.

Free-ranger
Chicken that lives in a free-ranging situation.

Frizzle
Heavy breed of chicken (see page 41).

Gamecock
Fighting rooster.

Gape Worms
Internal parasites (see page 169).

Gizzard
Part of the digestive system where food is ground up (see page 83).

Gizzard Worms
Internal parasites of chickens (see page 170).

Gold Brahma
Heavy, soft-feathered breed of chicken (see page 42).

Gold-laced Wyandotte
Heavy, soft-feathered breed of chicken (see page 36).

Gullet
See Oesophagus.

Hackles
Feathers on the sides and rear of a chicken's neck. Male hackles are usually more pointed than those of females.

Hamburg
Light, soft-feathered breed of chicken (see page 42).

Heavy breed
Breed of chicken which has a high meat-to-bone ratio and therefore is particularly suitable for eating.

Hock
Joint between the lower thigh and shank – occasionally, but incorrectly, known as the knee.

Houdan
Heavy breed of chicken (see page 43).

Hybrid
Result of crossing two pure-bred lines.

Hybrid vigour
A hybrid is said to have hybrid vigour, displaying the best and strongest characteristics of both parents.

Immunity
Ability to resist infection.

Incubation
Process of producing chicks from fertile eggs, either naturally with a mother hen sitting on them, or artificially by using a incubator.

Incubator
Piece of equipment that assists in the hatching of eggs.

Internal laying
Problem habit in chickens (see page 179).

Isthmus
The part of the oviduct where shell membranes are formed during development of an egg.

Japanese
Breed of true bantam (see page 59).

Jumping
Term occasionally used when a cock-bird mates with a hen.

Keel bone
Breast bone or sternum.

Lacing
Edging stripe around a feather; it differs in colour from the rest of the feather.

Layer
Hen currently laying eggs.

Leaker
Egg that leaks as a result of the shell being cracked.

Leg feathers
Feathers that protrude from the outer sides of the legs of some breeds of chicken, such as Brahmas and Cochins.

Leghorn
Light, soft-feathered breed of chicken (see page 43).

Lice
External parasites of chickens and other warm-blooded creatures (see page 168).

Light Sussex
Breed of chicken (see page 37).

Litter
Material used to cover the floors of chicken houses.

Magnum
Part of the oviduct that secretes a thick albumen during the process of egg formation.

Mandibles
Upper and lower parts of a beak.

Marking
General term for markings on plumage, such as barring, lacing, pencilling and spangling (see page 156 for details).

Minorca
Light breed of chicken (see page 44).

Mites
External parasites of chickens and other warm-blooded creatures (see page 166).

Morbidity
Percentage of chickens (and other creatures) affected by a disease.

Moult
Natural process of shedding old feathers and, later, growing new ones.

Mounting
Term describing when a cock-bird mates with a hen.

Muff
Cluster of feathers below and around the sides of the eyes and covering the earlobes.

Nankin
Breed of true bantam (see page 59).

Nest egg
Wooden, china or plastic egg placed in a nest or brooding box to encourage a hen to lay eggs.

New Hampshire Red
Heavy, soft-feathered breed of chicken (see page 44).

Notifiable diseases
Diseases that must be reported to animal welfare authorities (see page 165).

OEGB
Abbreviation for Old English Game Bantam (a breed of chicken).

Oesophagus (Gullet)
Narrow, muscular tube leading from the bird's mouth to its stomach.

Oil sac
Correctly known as the uropygial gland, this sac is found at the base of the tail. It assists a bird when preening or conditioning its feathers.

Old English Pheasant Fowl
Light breed of chicken (see page 45).

Onagadori
Light breed of chicken (see page 45).

Oven-ready
Describes a bird that has been plucked, eviscerated and prepared for cooking.

Oviduct
This is where eggs develop in a hen. It is formed of the funnel, magnum, isthmus, uterus and vagina.

Ovum (plural: ova)
Ova are round bodies attached to the ovary. They drop into the oviduct and become the yolks of eggs.

Parasite
Organism that lives on, or inside, a host

animal, deriving food and protection without giving anything in return.

Pasting
Loose droppings that stick to a chicken's vent area. This condition is also known as 'sticky bottom' and 'pasting up'.

Pathogen
Capable of causing a disease.

Pecking order
Natural order of seniority that evolves within a group of chickens (see page 16 for details).

Peking
Breed of true bantam (see page 60).

Pendulous crop
Crop which is enlarged and, sometimes, hanging down in an abnormal manner. Ageing hens often have this problem but, fortunately, it is not a serious one.

Perch
Horizontal pole with rounded edges where hens are able to rest and sleep.

Persistency of lay
Ability of a hen to lay eggs steadily and evenly over a long period.

Phenotype
General outward appearance of a bird, rather than its genetic make-up.

Pickout
This is when there is damage to the vent area in a chicken; usually, it is the result of cannibalism (see page 177).

Pin feathers
Short, undeveloped feathers that form short stubs.

Pin holes
Holes in shells of eggs that have been created by a bird's beak or claws.

Pinion
Tip of a bird's wing, from the last joint.

Pipping
Act of a chick breaking out of its shell.

Plumage
Feathers that make up the outer covering of a fowl.

Plymouth Rock
Breed of chicken (see page 46).

Point of lay (POL)
Age at which pullets start to lay eggs, usually between 20 and 22 weeks old. It is at this age that hens are often offered for sale by raisers of chickens.

Poland
Light breed of chicken (see page 46).

Pop hole
Doorway through which hens are able to enter and leave a chicken house.

Poultry
Generic term for species of birds that are used as food. These include chickens, ducks, waterfowl, turkeys, guinea fowl and quails.

Poussin
Small chicken, in live weight 0.5–1 kg (1–2 lb), sold especially for the gourmet market.

Precocity
When a pullet begins to lay eggs before being physically able to do so. This usually results in small, soft-shelled eggs.

Preening
Act of grooming and cleaning feathers.

Primaries
Long, stiff, flight feathers positioned at the outer tip of each wing.

Prolapse
Problem habit in hens (see page 179).

Pubic bones
Thin, terminal parts of the hip bones that form part of the pelvis. They are frequently used to judge the productivity of laying birds.

Pullet
Chicken less than one year old that has not started to lay eggs.

Rachis
Central, stiff and main part of a feather.

Range-fed
Describes chickens that are kept as foragers and allowed to wander and graze freely in a field.

Red Dorking
Heavy, soft-feathered breed of chicken (see page 47).

Redcap
Light breed of chicken (see page 47).

Reportable disease
See Notifiable disease.

Resistance
Immunity to infection from a disease.

Rhode Island Red
Heavy to medium soft-feathered breed of chicken (see page 48).

Roche scale
Scale used to measure the depth of colour in a yolk.

Roo
Shortening of rooster – mainly an American term.

Rooster
Male chicken (also known as a cock-bird or cock), mainly an American term.

Rose comb
Style of comb, flattened to the head, covered with small nodules and finished with a leader or spike.

Rosecomb
Breed of true bantam (see page 60).

Roundworms
Internal parasites of chickens and other poultry (see page 170).

Saddle
Lower part of the back, from the centre to the tail on cocks. In females, the same area is known as the cushion.

Scales
Small, hard, overlapping plates that cover a chicken's shanks and toes.

Scots Dumpy
Light breed of chicken (see page 49).

Scots Grey
Breed of chicken (see page 50).

Scratching
The habit of chickens to dig at the ground with their claws to find food, such as insects and grains.

Sebright
Breed of true bantam (see page 61).

Secondaries
Large, inner wing feathers adjacent to the body. They become visible when the wing is extended.

Self-colour
A breed where there is a single colour throughout, such as buff or white.

Set
To keep eggs warm, so that they will hatch.

Setting
Placing a group of eggs in an incubator or under a hen to encourage them to hatch. This is also known as sitting.

Sex-linked
Describes any inherited factor associated with the the genetics of either parent.

Sexual dimorphism
Differences between the average male and female birds in the same flock.

Shank
Lower part of the leg below the thigh.

Shell
Protective, semi-permeable casing, usually about 0.3 mm thick and almost entirely formed of calcium carbonate crystals, that encloses an embryo yolk and albumen.

Shell membranes
There are two shell membranes within an egg, forming a protective layer.

Sib
The resulting progeny from a brother-and-sister mating. Such matings are not recommended.

Sickles
Long, curved, top pair of feathers on the tails of some male birds.

Side-yolked
Where the yolk is offset within an egg.

Silver-laced Wyandotte
Heavy, soft-feathered breed (see page 36).

Silver Pencilled Wyandotte
Heavy, soft-feathered breed (see page 36).

Single comb
Comb with a single, upright blade.

Sitting
See Setting.

Spangling
Markings produced by a large spot of colour on each feather and differing from the overall ground colour.

Spanish
Breed of chicken (see page 50).

Spike
Pointed end of a rose comb and sometimes known as a leader.

Spur
Stiff, horn-like projection found on the legs of some birds. Spurs are positioned on the inner parts of the shanks (legs).

Started pullet
Pullet that has feathered out, but is yet to start laying eggs.

Straight run
This is when chicks are sold without first being sexed, resulting in male and female chicks being mixed together.

Strain
Variety that will produce the same characteristics when bred from one generation to another.

Sultan
Breed of chicken (see page 51).

Sumatra
Light breed of chicken (see page 51).

Tail coverts
Soft, curved feathers at the edges of the lower parts of some tails.

Tapeworms
Internal parasites of chickens and other fowls (see pages 170–171).

Testes
Male sex glands.

Thigh
Part of the leg above the shank.

Thread Worms
Internal parasites of chickens (see page 171).

Ticks
External parasites of chickens (see page 167).

Trachea
Windpipe, part of the respiratory system that conveys air from the larynx to the bronchi and lungs.

Treading
Sexual act of a mating.

Type
Shape and size of a chicken, defining its breed.

Variety
Subdivision within a breed.

Vent
External opening through which a hen expels an egg, as well as its droppings and urine.

Vet
Abbreviation for veterinary surgeon.

Vitelline membrane
Clear membrane that encases the yolk.

Wattles
Thin, pendant appendages at either side of the base of the upper throat and beak, usually much larger in males.

Welsummer
Light, soft-feathered breed of chicken (see page 52).

Yokohama
Light breed of chicken (see page 52).

Yolk
Central part of an egg that contains valuable nutrients (see page 99), but also, most importantly, houses the embryo chick and provides the means to a new life in a chicken's reproduction cycle.

Index